TECHNOLOGICAL
TRANSITION
IN CARTOGRAPHY

TECHNOLOGICAL TRANSITION IN CARTOGRAPHY

Mark Stephen Monmonier

THE UNIVERSITY OF WISCONSIN PRESS

Published 1985

The University of Wisconsin Press
114 North Murray Street
Madison, Wisconsin 53715

The University of Wisconsin Press, Ltd.
1 Gower Street
London WC1E 6HA, England

Copyright © 1985
The Board of Regents of the University of Wisconsin System
All rights reserved

First printing

Printed in the United States of America

For LC CIP information see the colophon

ISBN 0-299-10070-7

FOR MY PARENTS
John Carroll Monmonier
and Martha Mason Monmonier

Contents

Illustrations

Preface

This book was conceived to explain to students why my introductory course in cartography pays little attention to drafting maps with pen and ink. Many undergraduate students have heard that people skilled in drawing maps have found jobs even in tight recessions, and they become worried when laboratory exercises focus on critiquing design and using computers rather than inking lines and lettering labels. For these readers the book should offer insight on the extent to which computers and other electronic technology will alter the form of the map, increase map use, and reduce the skill and training required to obtain a decent-looking, convincing cartographic display. Because of what I call the Electronic Transition in Cartography, the ability merely to plot a "good looking" map will become as commonplace as the ability to type a "good looking" letter. The good student ought not be discouraged by a lack of drawing ability, and the forward-looking student must plan for change and appreciate the role in mapping of public policy and management as well as design. Cartography, after all, is no more drafting than journalism is typing or chemistry is cooking.

In these essays I explore the changes that electronic technology has made, is making, and probably will make in all principal phases of mapping and map use: navigation, surveying, land information systems, decision support systems, and map publishing. I examine previous technological transitions to assess the extent to which significant inventions can radically transform the content, appearance, use, and worth of maps. Cartography's Electronic Transition will render the map as much the outcome of a political process as it traditionally has been the product of exploration and craftsmanship. The cartographer, the geographer, the social scientist interested in regional patterns, and the earth scientist must be aware of the complex linkages among the many producers, distributors, and users of geographic information. This book is intended to provide much of the general technical background needed to appreciate the range of problems that mapping policy ultimately must confront.

The book should also interest educators and humanists. It predicts that the digital map will replace the paper map as the principal medium of cartographic storage and analysis, and it forecasts an era in which the immediate user rather than a distant map author/publisher will compose most cartographic displays. Yet it also recognizes that modern telecommunications could foster an increased appreciation and use of maps by journalists and geographers, and thus expand the market for cogent, well-planned presentational cartography. But whether an increased exploitation of the map will serve knowledge and understanding more than hype and glitter is uncertain, for without a graphically literate and geographically astute public, maps may become little more than pleasurable icons employed to seduce and entertain rather than to inform and enlighten.

At a time of increased concern for human survival, parts of this book might anger some readers, as they did one reader of the manuscript who was alienated by my frequent and sometimes enthusiastic mention of military contributions to cartographic innovation. But whatever our politics and social conscience, we cannot deny the strong historical bond between map and soldier, nor can we dispute that much of the present digital cartographic effort is inspired or sustained by a concern for national defense. I cannot convincingly explain and condemn the cruise missile in the same breath, and I choose not to weave into these chapters the uncertain and highly pessimistic threads of my personal doubts about the ability of the world's governments to resist the temptations and threats of nuclear war.

I owe several people a debt of gratitude. Guthrie Birkhead, Dean of the Maxwell School of Citizenship and Public Affairs at Syracuse University, granted me a semester leave to plan and organize this book. Many colleagues graciously read and provided helpful comments on early drafts of individual chapters: Gregory Bass, Lee Bender, John Bossler, Vinton Cerf, Richard Dahlberg, Andrew Douglas, John Jensen, Jon Leverenz, Robert McMaster, Judy Olson, Timothy Petersen, George Schnell, Anthony Williams, and David Woodward. Michael Kirchoff, Maureen Devey, Andrew Bratton, Michelle Kermes, Ann Perry, Marcia Harrington, Sean Cassidy, David Flinn, and Janet Saxon at the Syracuse University Cartographic Laboratory prepared the artwork for most of the illustrations. Michael Peterson, Joel Morrison, Jon Kimerling, and Barbara Petchenik provided useful suggestions on the complete manuscript. David Woodward was a source of much appreciated encouragement and counsel. My wife Margaret gave me the tolerance and nourishment that makes writing both possible and purposeful.

TECHNOLOGICAL
TRANSITION
IN CARTOGRAPHY

CHAPTER 1

Introduction

This book is about technological innovation and maps. It is in part a cursory technological history and in part a philosophical essay on the evolution of maps, map information, and map use. It examines change in the form and function of maps, with a focus on the twentieth century, and probes the extent of change likely by the early twenty-first century in the variety of uses of map data and the ways in which these data are collected, stored, disseminated, displayed, and analyzed. The current technological upheaval can be called the Electronic Transition after the physical agent for coding, storing, and transmitting geographic data processed by the digital computer. Other, related innovations of modern electronic technology such as orbiting satellites, fiber optics, and silicon-chip memory are inexorably involved in this revolution in the form, use, value, and values of maps.

The Electronic Transition is well underway. By the mid-1970s the analysis of geographic information stored on magnetic tape or disc was sufficiently prominent to foster a new term, *digital map*. In its *Glossary of Terms in Computer-assisted Cartography,* the International Cartographic Association (ICA) is lexicographically oblique in its definition of the digital map as "a term sometimes loosely used for '*digital* cartographic *data*'" but more informative as well as mildly philosophical in its definition of "digital" alone:

> Pertaining to *data* in the form of *digits*. Contrast with "*analog.*" Besides representing *numbers* (quantities) the digits are also used

to form *codes* which represent *characters* out of a finite set. Therefore a *data representation* making use of a finite set of characters is
already called "digital" (e.g. text), and there is a slight difference
between "digital" and "numerical," which are synonymous only when
numerals are used to represent characters.[1]

The ICA Glossary also includes definitions for digital cartographic model, digital image, digital manuscript level, digital
manuscript tape, digital model, digital mapping system, digital
plotter, and digital tablet, indicating that the concept of digital map already has a wide base. The fourth edition of Arthur
Robinson's widely used textbook *Elements of Cartography* provides a more concise definition of digital as "the representation of
a quantity by a number code; contrasted with analog."[2] Thus a
map may be analog or digital, and hence either graphic or un-
seen—by human eyes, that is. A computer can examine a digital
map and from it produce an analog map. Digital maps also can be
sent rapidly over thousands of kilometers by wire or electromagnetic waves. Because digital data are readily changed, the map
information thus forwarded should also be up-to-date as well.
Continued advances in computers and telecommunications will
assure an even fuller development of digital cartography and alter
attitudes toward maps and their use.

The paper map might well be one of the casualties of cartography's Electronic Transition. Little more than a decade ago almost
every map was conveniently available only on paper. If a user
wanted to examine a copy, he viewed a paper copy. If a customer
ordered a copy, he was sent a paper copy. Except for plastic or
glass drafting and production materials, microform copies, globes,
a few maps printed on silk for downed aviators or on plastic as
instructional exhibits, and cognitive maps in people's minds, all
maps existed on paper. In the decades ahead, though, the digital
map will displace the paper map from its dominant position. Although a copy on paper or some other permanent graphic reproduction medium should be obtainable for almost all maps, paper
will no longer be the principal medium for compiling, storing,
distributing, displaying, and viewing geographic data. Moreover,
most paper maps will be affected in their design or selection by
digital cartography. Before deciding upon a set of colors and sym-

bols, a map's designer probably will preview on a cathode ray tube (CRT) display screen numerous symbolic representations of the desired distribution. Before requesting a printed copy of that particular map, the user sitting in front of a similar display terminal might view hundreds of potentially interesting distributions. The computer data base, not the paper map, will be the ascendant cartographic form from which most other map copies and map analyses will be derived.

A Century of Cartographic Change

How much progress is likely in how short a time? Is a prediction of the dominance of the digital map early in the twenty-first century not perhaps looking ahead too far too soon? An examination of past cartographic progress might relieve some of the skeptical reader's doubts about the rate of improvement possible in maps and mapping technology. The following two examples, each covering approximately a century, illustrate the significant recent change in both geographic detail and the richness of map information.

The first example focuses on the village of Manchester, Vermont, a popular resort in the Taconic Mountains. One of the most detailed early maps of the area is the portion shown in Figure 1.1 of a sheet in the *Atlas of Bennington County, Vermont,* published in 1869.[3] The original scale of this map is approximately 1:48,000. Its symbols and type emphasize political units, the transport system, and individual residences. Surveying methods were crude, and many atlases of this cartographic genre were hastily produced advertising vehicles replete with self-flattering engravings of the residences of those landed gentry who were persuaded to buy subscriptions from the atlas publisher's field corps of salesman-surveyors.[4]

The publication in 1894 of the Equinox, Vermont 15-minute topographic quadrangle map marked a significant advance. Figure 1.2 shows the Manchester portion of this map sheet. With a scale of 1:62,500 and a contour interval of 20 feet [6 m], the Equinox sheet was part of the U.S. Geological Survey's attempt to provide an accurate, systematic, standardized "mother map" of the nation's terrain, hydrography, political boundaries, transport net-

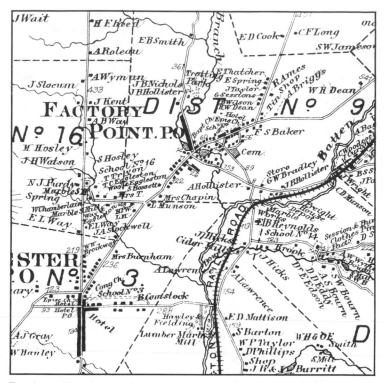

Fig. 1.1.   The vicinity of Factory Point, Vermont, later called Manchester, as
shown in the *Atlas of Bennington County, Vermont,* published in 1869. This portion of a larger map has been reduced photographically from a scale of approximately 1:48,000 and formatted to match Figures 1.2 and 1.3 in scale and areal
extent.

Source: F. W. Beers, G. P. Sanford, and others, *Atlas of Bennington County,
Vermont* (New York: F. W. Beers, 1869), plate 13. Copied from the collection of
the Library of Congress.

work, landmarks, and place names. Progress was slow, and coverage was not complete for Vermont until the 1950s. But by that
time the Geological Survey had undertaken an even more ambitious project, its 7.5-minute topographic series with four, more
detailed 1:24,000-scale maps covering the same area represented
on a single sheet at 1:62,500 in the old 15-minute series. The
Manchester 7.5-minute sheet was not published until 1968. Aside

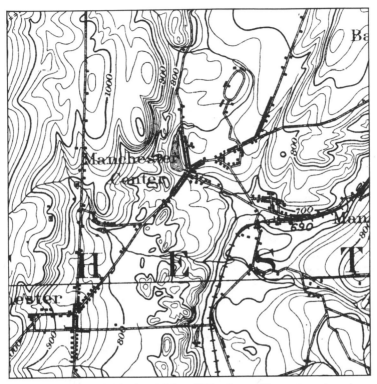

FIG. 1.2. Topography in the vicinity of Manchester, Vermont, as shown on a portion of the Equinox, Vermont U.S. Geological Survey 15-minute quadrangle map, 1894. Map has been enlarged from original scale of 1:62,500. Area shown is a square approximately 3.9 km [2.4 mi] on a side.

from a green area pattern showing the distribution of woodland, the most noteworthy changes were the increased accuracy of the contours and horizontal positions and the revision of the expanded network of village streets and town roads. Figure 1.3 shows transportation routes and other cultural features from the 7.5-minute sheet.

Improved accuracy, increased detail, and more current cultural information are not the only hallmarks of cartographic progress represented in Figure 1.3. More significant than the updated landmarks and revised road network are the boundaries and numeric

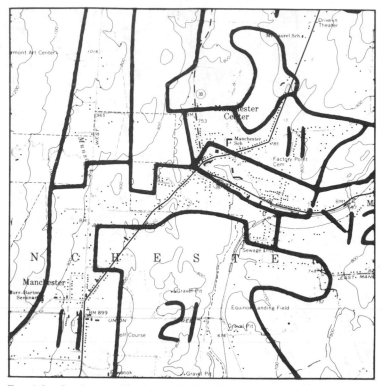

FIG. 1.3.   Land use and land cover in the vicinity of Manchester, Vermont, as
shown on a portion of the Glens Falls, New York, U.S. Geological Survey Land
Use and Land Cover Map, no. L-185, 1972. Map has been enlarged from original scale of 1:250,000 and superimposed upon features from the Manchester,
Vermont 7.5-minute U.S. Geological Survey topographic quadrangle map, 1968.
Topographic map has been reduced from original scale of 1:24,000, and hydrography and some contours have been suppressed. Land categories labeled
include Residential land (11), Commercial and Services (12, to the right of the
center of the figure), and Cropland and Pasture (21). Closed polygon in the upper
right is Evergreen Forest land. Land categories around the perimeter of the figure, starting at the lower left and proceeding clockwise, are Residential (11),
Mixed Forest land, Cropland and Pasture, Evergreen Forest land, Cropland and
Pasture, Evergreen Forest, Mixed Forest, and Cropland and Pasture (21).

codes representing categories of land use and land cover. These symbols are bold in this example because they have been enlarged and transferred photographically from the Geological Survey's 1:250,000-scale Glens Falls, New York, Land Use and Land Cover map sheet, which covers a quadrangle encompassing one degree of latitude by two degrees of longitude. Although larger and more generalized, this sheet is far richer in thematic detail than a standard topographic map. It is also significant as part of a systematic attempt to map land use and land cover according to 37 categories for the entire lower 48 states and Hawaii.[5] Furthermore, this land-classification coverage is only one part of a digital geographic data base that includes representations of drainage basins, political units, census reporting areas, and federal land ownership.[6] Begun in the early 1970s, national coverage should be complete—and ready for revision—in the late 1980s.

Were this book printed in color, more visually dramatic examples of cartographic progress could be presented. For the reader with a personal graphic display system, it would be possible also to demonstrate that no longer need cartographers think of maps largely as "sheets," bounded by fixed rectangles and joined to cover larger areas only with difficulty. Digital cartography can readily provide displays for irregular areas and regions heretofore divided among several adjoining quadrangles.

The Commonwealth of Pennsylvania is the focus of the second case study, an examination of copyrighted maps, produced principally by private firms and individuals. Table 1.1 shows percentage frequencies of copyrights by map type for three five-year periods spanning a century of progress in commercial, private-sector cartography.[7] The differences between periods are noteworthy. The early 1870s were marked by the county atlas, the state outline map, and the mineral survey; the early 1920s were dominated by highly detailed fire underwriters maps for urban areas; and the early 1970s were marked by city street guides and various road maps. Particularly significant, though, is the absence for 1920–24 and 1970–74 of the county atlas, and the absence in 1970–74 of fire insurance maps. Needs and capabilities change, and a half century can alter radically both the mix of maps and the dominant type.

TABLE 1.1. Variety of maps for the Commonwealth of Pennsylvania and places therein copyrighted 1870–1874, 1920–1924, and 1970–1974.

Type of Map	Percentage Frequency		
	1870–74	1920–24	1970–74
County atlas	21		
State atlas	18	2	
Mineral	13	2	
City atlas	8		
City map	8	8	56
County topographic	5		
Facsimile or historic	5		1
Fire insurance		58	
State road		6	12
County road		10	8
State wall or folded		4	
Tourism		3	5
Utility and railroad		3	4
Local engineering/planning		2	1
County outline		1	
Advertising market area			3
Recreational			4
Urban transit			1
Other	22	1	5

Source: Generalized from Tables 1, 2, and 3 in Mark S. Monmonier, "Private-sector Mapping of Pennsylvania: A Selective Cartographic History for 1870 to 1974," Proceedings of the Pennsylvania Academy of Science 55 (1981): 69–74.

The apparent increase in the variety of maps after the 1870s had its roots in the first half of the nineteenth century, and a bit before, when numerous technical and conceptual innovations spawned many new symbols, especially for thematic maps. As carto-graphic historian Arthur Robinson has carefully documented, the two hundred years between the mid-seventeenth and mid-nineteenth centuries witnessed a major revolution in cartography. Indeed, this period produced most of the symbols now employed widely on statistical maps and in atlases.[8] Cartography has seen other revo-lutions as well, most notably that following the development of printing in the fifteenth century. The printing revolution increased the number of maps in circulation, and the thematic revolution increased their information content. The Electronic Transition, which mapping has recently entered, should bring changes as deep and broad as these previous upheavals.

Mapping and the Rate of Technological Progress

Astute cartographers have long been aware of the accelerating development of mapping technology. In 1937, for example, Erwin Raisz published an illuminating, compact graphic summary of cartographic innovations.[9] Raisz's paper includes six full quarto-size pages with the names of innovations and innovators plotted against a vertical axis representing time. His first chart, Antiquity, covers the nine and a half centuries from 600 B.C. to 350 A.D. and mentions 31 separate persons or concepts. His last two pages, labeled Modern, cover the 230 years between 1700 and 1930, and mention over 250 separate persons, concepts, or events. Equal amounts of space accommodate four decades of Antiquity and a single decade of the Modern period. Were Raisz alive today to extend his "Chart of Historical Cartography" into the 1980s, a comparable level of detail might well require an extension four pages wide to cover but a single year in each vertical centimeter.

A full appreciation of the Electronic Transition and its probable effects must recognize the increased integration of mapping with other, larger technologies such as computer hardware, telecommunications, and solid state physics. The accelerated pace of development in these areas reflects the society's resources and its attitudes, objectives, resources, and technical abilities. Cultures grow and become more complex, thereby establishing new needs that in turn lead to new priorities. These new priorities also create new opportunities, with high profits for successful innovators.[10] The resources made available for research and development also are important. Because of its value to both military defense and economic development, mapping has received an especially abundant share of research and development resources.

John Lienhard's concise graphic summary of technological progress illustrates the explosive growth in science and engineering underlying the accelerated development of mapping technology.[11] Lienhard calculated a time-constant of change for 14 innovations, from the mechanical clock to the printed circuit. His time-constant is roughly the number of years needed for a technology to improve about 2.71828 times—he called this one "*e*-folding." It is based on the theory of exponential growth and the mathematical constant *e* (2.71828...), the base of the natural logarithms. Lienhard plotted this comparative measure of technological progress against the date of each innovation's inception. A significant shift was noted after 1832, as shown in Figure 1.4. From 1400 until well into the nineteenth century, about 40 years was required for one *e*-folding. As shown by the generalized curve in Figure 1.4, the time-constant has declined steadily since 1832, leveling off somewhat in the mid-twentieth century. This curve illustrates continued progress in the overall rate of technological development, with more rapid improvement for such newer technologies

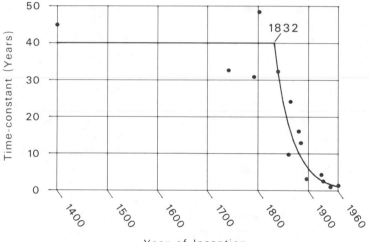

FIG. 1.4.   Increased rate of technological improvement as illustrated by generalized relationship between year of inception of a technology and Lienhard's time-constant of change—roughly the number of years required for the technology to improve about 2.7 times. Dots represent various technologies upon which Lienhard's study is based.

Source: Plotted from formula and data reported in John H. Lienhard, "The Rate of Technological Improvement Before and After the 1930s," *Technology and Culture* 20 (1979): 515–30. Lienhard's original graph portrayed a log-linear relationship.

as digital computers, particle accelerators, liquid-fuel rockets, and printed circuits.

Technological historian Donald Cardwell attributes this accelerated progress to the evolution of a class of professional scientists and the establishment of research institutes.[12] To be sure, today's scientists and engineers are quite different from the largely self-taught amateurs and entrepreneurs of the nineteenth century. He identified four stages in the development of technology, beginning with the Middle Ages. Cardwell's last phase began about 1860 with the advent of the industrial research laboratory. He aptly quotes philosopher Alfred North Whitehead's observation that "the greatest invention of the nineteenth century was the invention of the method of inventions." Government and university research programs also marshaled the personnel, information base, and management structure required for rapid technological advance. Mapping, a field long allied with the military as well as with civilian public works agencies, had an institutional foundation that fostered its participation in technological progress.

Adapting and Institutions

In the late twentieth century, as indeed throughout much of its earlier history, cartography has been less a field of original innovation than one of clever adaptation and application. This view in no way degrades the originality of recognizing an application to geographic measurement or representation of some new or long overlooked principle of mathematics, engineering, art, or polymer chemistry. Progress through secondary or transfer innovation is perhaps inevitable given the small number of cartographers in comparison to the much larger number of engineers and physical scientists with at least a peripheral interest in things that might be employed in mapping. Thus, invention of the method of technological invention, as observed by Whitehead, is not sufficient for progress in cartography: innovation is needed as well in the organization and management of programs for the collection and dissemination of geographic information. If cartography is to continue to advance, its institutions as well as its researchers, operatives, and theories must also advance.

In a recent monograph examining diverse approaches to the history of cartography, Blakemore and Harley suggest that maps might best be studied within the conceptual framework of a "living language."[13] Modern technology suggests with equal vigor that mapping—the process more than the image—should also be a focus, to be examined through its living institutions, and their economics, geography, politics, and sociology. Cartographic scholars must develop for public policy, government agencies, the private sector, the professional press, and the professional organization the same scrutiny they have lavished upon graphic symbology, map design, data collection, and other superficially more enigmatic problems. If the fifteenth century saw a revolution in publishing and the nineteenth century a revolution in design, the late twentieth and early twenty-first centuries must see a revolution in organization and management.

The pages that follow should heighten awareness of the inevitability of change in maps and mapping among policy makers, map producers, map users, educators, and students. The next chapter, for instance, in explaining the role maps have played in navigation, notes that navigation is now possible without analog maps, but not without new cartographic products and new demands upon cartographers. Mapping's future will precipitate new problems and societal expectations, and will demand new skills and management strategies. Exactly when that future will arrive is, of course, moot, yet there can be little doubt that mapping and geography are being propelled forward by the technological currents in which they have become engulfed.[14] What is not at all certain is whether maps will be better designed and more accessible, and whether society will obtain the best return for its investment in mapping infrastructure. Successful, economical adaptation is cartography's foremost challenge.

CHAPTER 2

Location and Navigation

Computers, modern telecommunications, and orbiting satellites are changing the form of maps and their role in navigation. Although helpful in many ways as a tool for travel, paper maps are by no means indispensable to the navigator of the late twentieth century. A ship or aircraft can readily determine its position without a map, and its crew can plan and follow a course without consulting a visual display. Yet when maps are employed, they can be more up-to-date, richer in useful information, more accurate, and more easily stored, retrieved, and handled. Where humans do not abdicate to autopilots, maps can be ever more powerful tools.

This chapter traces the role of the map as an aid to navigation and shows how technological innovations and scientific discoveries in astronomy, instrumentation and measurement, cartography, military and naval science, and telecommunications have improved our ability to select an appropriate course and follow it. Navigation is perhaps the most suitable theme for this first chapter after the introduction, for most early maps were produced to assist travelers in finding their way in unfamiliar territory.

Map Use Tasks in Navigation

Navigation has two principal phases, route planning and route following. Figure 2.1 illustrates the map use tasks in route planning. The navigator must first determine the locations of his pres-

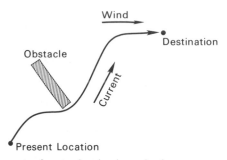

FIG. 2.1.   Elements of route planning in navigation.

ent position and his destination. He must then select a path from
origin to destination that avoids such obstacles as coastal irregu-
larities or a submerged reef. The map must show intervening
barriers and hazards as well as locational information. Lines of
ready travel—highways, railroads, ocean currents, and prevailing
winds—should also be represented. Some courses minimize travel
time, whereas other routes minimize distance and fuel consump-
tion. The navigator uses a variety of geographic distributions in
choosing an optimal course.

In the route-following phase, progress along the course is mon-
itored by comparing one's current location to the location of the
intended path. In land navigation or when flying at a low or medium altitude on a clear day, this comparison can be based upon
visual signposts, or landmarks. The map is a convenient device
for storing the locations of significant terrain features and for
retrieving information about landmarks. In contrast, when sailing
or flying at night or in cloudy weather, visual landmarks are of
little use, and other guides must be used if certain progress is to
be made. Radar and Sonar can help the navigator avoid nearby
hazards. Radio beacons serve as signposts for "sightless" navigation—their signals are readily associated with mapped positions,
and with simple triangulation the navigator can recover his location.

In addition to these nearby yet invisible landmarks, the navigator has often depended upon navigational signposts light years
away from Earth. Before the invention of radio, navigators at sea
relied on cosmic beacons—the Sun and other stars, the planets,

and their satellites. Cosmic "signals" in the form of relative position, eclipses, and elevation above the horizon were often read with uncertain precision. Tedious calculations translated imprecise measurements into marginally accurate locations. Today, artificial satellites broadcast time signals to on-board receivers with built-in computers, and the navigator can rapidly determine his position to within several meters.

Orientation, Wave Crests, and Stick Charts

All locations are relative, but some reference systems are more nearly absolute or obvious than others. A city street network laid out as a rectangular grid facilitates two-dimensional coordinate referencing if, for instance, north-south streets are lettered from west to east as A, B, C, and so forth, and if east-west streets are labeled First, Second, Third, and so on. Thus the intersection of H street and 5th Avenue is readily located seven blocks east and four blocks north of an origin at A Street and 1st Avenue. Less systematic are such directional references as "left turn at the red barn" or "fifty kilometers southeast of Pittsburgh."

Locational references may be globally significant, if based upon a systematic standard, recognized throughout the world. References may also be only locally meaningful if based upon idiosyncrasies of the immediate neighborhood. The "left turn" is a purely local reference, based on a locally significant direction of travel, whereas the "southeast" descriptor is a mixed reference, based on both a local reference point, Pittsburgh, and the globally meaningful cardinal directions "south" and "east." The urban street grid may be aligned either to the globally significant true North or the local Magnetic North. Many rectangular street patterns were based upon the misleading convenience of Magnetic North—a convenience diminished by the wandering of the North Pole about the Arctic.[1] Other idiosyncratic street grids are aligned to a coast or river bank, as in The Dalles, Oregon, or to linear topography, as in State College, Pennsylvania (Figure 2.2).

The consistent orientation of wave crests, a pattern produced by prevailing winds, provides primitive navigators with a locally meaningful directional reference. The Marshall Islanders of the

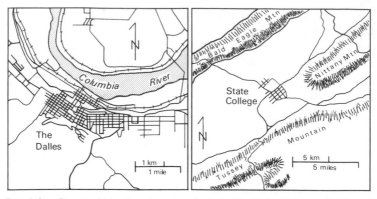

FIG. 2.2.   Street grid is aligned to river bank in The Dalles, Oregon (left), and to nearby northeast-southwest trending ridges in State College, Pennsylvania (right).

South Pacific represent the trends of wave crests with sticks of palm ribs. Held together by palm fibers, these stick charts are useful at sea as well as along the shore (Figure 2.3). The charts generally are about 40 to 60 cm [16 to 24 in.] square, with small shells or pieces of coral marking the locations of islands.[2]

These crude maps are useful because an island's neighbors often are not visible above the wave swells on the horizon. The atolls of the Marshall Islands seldom reach an elevation greater than 6 m [20 ft]. Neighboring islands 50 km [31 mi] away are not visible above the rough sea. Wave fronts striking an atoll and its submerged base are reflected and refracted to produce a choppy, amplified interference pattern (Figure 2.4). Native boatmen rely upon these cues, which they can sense more easily through the nerves in their backs by lying in the bottom of the boat than by observing the water surface.[3] Primary swells may strike an atoll from more than one direction thereby providing additional cues.

Experienced native navigators have a well-developed mental map of the spatial arrangement of wave crests and islands, and do not need to rely upon stick charts.[4] These primitive Marshallese maps would seem to serve principally as training aids or safeguards against loss of memory. Although useful as physical models for teaching the principles and patterns of wave refraction, these charts are carried by the sailors in their boats, and no doubt

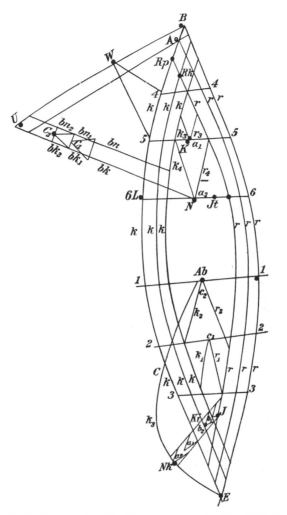

FIG. 2.3.   Scale diagram of a rebbelib, or group chart, of the Ralik chain in the Marshall Islands. The capital letters identify mussel shells representing individual islands.

Source: Captain Winkler, "On Sea Charts Formerly Used in the Marshall Islands, with Notices of the Navigation of These Islanders in General," *Annual Report of the Board of Regents of the Smithsonian Institution for the Year Ending June 30, 1899* (Washington: Government Printing Office, 1901), pp. 487–504 [ref. p. 501].

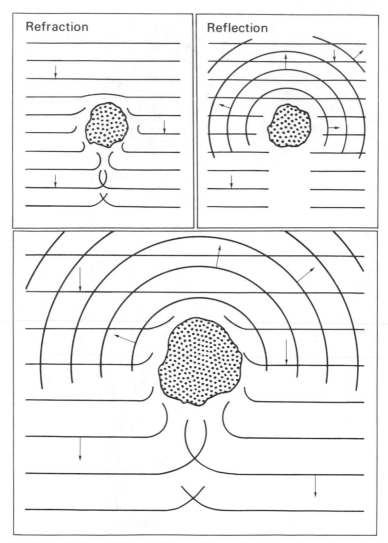

FIG. 2.4.   Atoll struck by a set of generally parallel primary wave crests, which
are both refracted (upper left) and reflected (upper right), produces an interference pattern of primary and secondary wave crests, which serve as navigation
cues for boatmen in the Marshall Islands.

play the same role in route following as the local road map in the glove box of the family car.

Pilot Charts, Map Grids, and Mercator's Projection

Until the fourteenth century, navigator's maps were more akin to the primitive Marshallese stick charts than to the modern hydrographic chart. Medieval European navigators used pilot charts called *portolani,* with detailed representations of coastlines.[5] The principal symbols were *rhumb lines,* each representing a constant sailing direction under a steady wind. Each chart had several wind roses, from which rhumb lines radiated in 8 or 16 equally spaced directions (Figure 2.5). Relative distances were not greatly distorted, and identifiable land points provided the basic frame of reference.

Navigation was by piloting, noting progress relative to observable landmarks, and by "dead" reckoning, that is, by *ded*ucing the ship's location from the directions and distances traveled. Primitive instruments enabled the navigator to estimate speed and distance traveled. A log tied to a coil of knotted line was periodically cast adrift.[6] The ship's velocity was computed from the number of evenly spaced knots pulled into the water during the time required for the sand to drain from the upper to the lower half of a sand glass. A traverse board, with pegs inserted in holes to indicate direction as well as distance, provided a record of progress. Pilot charts were used both for following a route along the coast and for planning a route to be followed by dead reckoning between opposite shores.

The linking of direction, the magnetic needle, and the Earth's magnetic field was a major advance in marine instrumentation. A magnetic needle can show an approximate north-south direction during the day or at night, and in fog as well as in clear weather. The magnetic needle was probably discovered in China and introduced into Europe during the Dark Ages by the Arabs.[7] In 1269, Petrus Peregrinus, an engineer with the French army, mounted a magnetized needle on a pivot surrounded by a circle marked for measuring angles.[8] Although Peregrinus could easily estimate magnetic bearings, the traditional mariner's compass was not "in-

Fig. 2.5.   Portion of a portolan chart prepared in 1529 by Diego Ribero, Spain's Royal Cosmographer. Map shows eastern United States and direction lines converging at compass roses.

Source: Edward B. Matthews, "The Maps and Map-Makers of Maryland." In *Maryland Geological Survey, Volume Two* (Baltimore: The Johns Hopkins Press, 1898), pp. 337–488 [ref. p. 344].

vented" until 1302, when the graduated circle was replaced by the wind rose. Navigators could then refer to the eight principal winds, their eight additional subdivisions called "half winds," and sixteen further subdivisions, called "quarter winds." With these early compasses, mariners could estimate direction, relative to Magnetic North, to the nearest of 32 points on the compass card.

Like other improved instruments for sailing, the magnetic compass increased both marine commerce and the utility of navigation maps.[9] Ships could navigate with greater assurance at night and through cloudy weather, marine employment became more stable, investment in ships increased, and voyages became more frequent. Relative positions of coastal features could be described more precisely, and sailing charts, used now more than ever for navigating directly across the Mediterranean rather than along its coast, became more detailed.

Translating an intended course into a compass direction remained a major difficulty. Although North was at the top and distances were not greatly distorted, the earliest portolan charts lacked a grid of meridians and parallels and were not based upon systematic, mathematically derived map projections. Later sailing charts based upon grids of uniformly spaced, parallel, straight-line meridians and parallels distorted area without making life easier for the navigator. Because the Earth's meridians converged toward the poles, the square grid also distorted angles by an east-west stretching that became more and more pronounced with increased distance from the equator (Figure 2.6). For a place on the

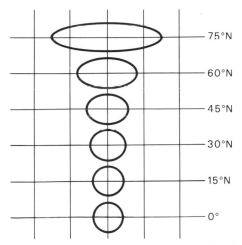

FIG. 2.6.   With increased distance from the equator, the distortion of angles on a square grid increases as increased east-west stretching distorts circles into ellipses.

30th parallel, a bearing might be distorted as much as 8 degrees. At the 60th parallel, some bearings are distorted as much as 20 degrees.

Frequent course corrections were necessary because direction would shift gradually for a course plotted on the map as a straight line. Before precise chronometers permitted accurate estimates of longitude, many navigators practiced *parallel sailing,* by traveling due north or south to the destination's latitude, and then east or west along the parallel by dead reckoning to the intended longitude.[10] Time and distance were sacrificed for simplicity and reliability.

Gerhard Mercator, a Flemish designer and manufacturer of maps, globes, and instruments, provided a solution. The projection bearing his name not only represents lines of constant direction as straight lines but also permits the navigator to read the true geographic bearing directly from the map as the angle between meridian and rhumb line. His solution was simple in principle—vary the spacing of the parallels to straighten all rhumb lines (Figure 2.7). Straightening the rhumb lines requires a steady increase in the separation of the parallels because a rhumb line, unless directly east-west or north-south, would spiral around and around

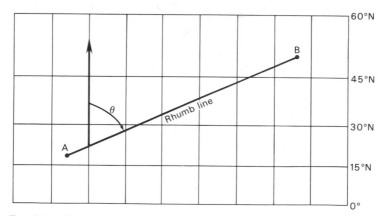

FIG. 2.7.   On the equatorially centered Mercator projection, parallels are spaced progressively farther apart with increased distance from the equator so that a straight line between points A and B represents a line of constant bearing called a rhumb line. The bearing angle, theta, can be measured directly from the map.

on the Earth as it proceeds poleward. The poles hypothetically would be straight lines, parallel to the parallels, with an indefinitely large east-west scale and located an infinite distance from the equator. Hence polar areas are seldom portrayed on a Mercator projection centered on the equator.

Mercator's original projection, as laid out in 1569 with mechanical drawing instruments, was of little value to navigators.[11] Because of the continual poleward increase in scale, unless a sailor was on one of the parallels shown on the map, he had no reliable means for comparing his actual location with his intended location on a rhumb line. As a result, the straight-line projection was little used until 1599, when Edward Wright, a professor of mathematics at Cambridge, published a graphic and mathematical explanation of the projection. Most important for navigators, Wright provided formulas and a table that could be used to compensate for the scale variation between two neighboring parallels. Mariners could then relate any map position to latitude and longitude, and Mercator's projection became an accepted aid to navigation.

The gnomonic projection is a useful companion to the Mercator projection. Whereas a straight line on the Mercator represents a constant direction, a straight line on the gnomonic represents the shortest distance between two points on the Earth. The two projections can be used in concert, with the shortest-distance, great-circle route divided into multiple segments the junctions of which are transferred from the gnomonic to the Mercator (Figure 2.8).

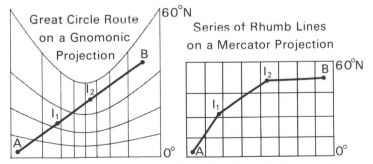

FIG. 2.8. A great circle route A-I_1-I_2-B, plotted as a simple straight line on a gnomonic projection (left), can be approximated by a series of rhumb lines on a Mercator projection. The great circle route minimizes distance traveled, but the series of rhumb lines is more easily navigated.

The former projection permits the navigator to plan an efficient course, whereas the latter provides him with the sailing directions needed to follow that course.

Thales of Miletus (640–546 B.C.), the founder of Greek geometry, developed the gnomonic projection over two thousand years before Mercator.[12] Although much older, the gnomonic was little used until Wright perfected the Mercator projection toward the end of the sixteenth century. Nonetheless, "great circle sailing," by following short rhumb-line segments along a great circle, was not commonplace until the nineteenth century. A principal difficulty was the tedious and often uncertain estimation of a ship's longitude.

Latitude and Longitude, Angles and Time

Latitude is the more "natural" of the Earth's two spherical coordinates and can be measured with relative ease. Longitude is arbitrary and can be determined only with sophisticated instruments or tables. The Earth has only two natural reference points, the North and South Poles, where it is pierced by its axis of rotation. Another, derived reference standard is the equator, the locus of points on the sphere midway between the poles. Strictly speaking, though, the Earth is not a perfect sphere, and the equator is defined more precisely as the intersection of the Earth's surface and a plane passing through the Earth's center and perpendicular to its axis. Through each point anywhere on the Earth between the equator and the poles there passes one parallel, a small circle also formed by the intersection of the Earth's surface and a plane perpendicular to the axis of rotation (Figure 2.9, left). The parallels are aptly named: their planes are parallel to the plane of the equator. An individual parallel is referenced by its latitude, that is, its angular distance north or south of the equator.

The meridians provide a convenient second coordinate, useful for differentiating among locations on the same parallel, but they have no natural origin or anchor. Except at the poles, there extends through each point on the Earth a half great circle formed by the intersection of the Earth's surface with a plane through the axis of rotation (Figure 2.9, right). These meridians run from pole to

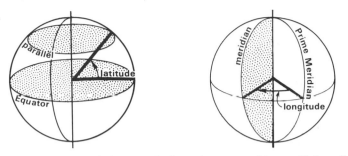

FIG. 2.9.   Latitude (left) is the angle above the equator of the specified parallel, and longitude (right) is the angle between the prime meridian and the specified meridian.

pole, converging at the poles and most widely separated at the equator. Individual meridians can be identified by their longitudes, their angles east or west of some arbitrarily chosen reference meridian.

A well-known, widely accepted point on the Earth's surface is the best anchor for a reference meridian. The geographer Claudius Ptolemy (c.100–c.170 A.D.) used a prime meridian through the westernmost known point, the Fortunate Isles, now called the Canary Isles, off the west coast of Africa.[13] All meridians on his maps were represented by longitudes measured in a single, positive, eastward direction. Global exploration doomed the Canaries as a natural anchor when Columbus rediscovered and explored America in the late fifteenth and sixteenth centuries.

A coordinate should be measurable as well as unambiguous, and because of the Earth's rotation, latitude is both. The Sun provides the most direct measurement of latitude, which is related to the *zenith distance,* the angle of the Sun from a point directly overhead (Figure 2.10, left). Zenith distance varies with time of day and is at its least at noon. Because the Earth's axis of rotation is inclined 23½ degrees from a hypothetical "axis of revolution" about the Sun, the zenith distance at noon also varies throughout the year. For places north of the Tropic of Cancer, the parallel at 23½ degrees north, the Sun is never directly overhead at noon. The smallest zenith distance occurs at noon on the summer solstice, and the largest zenith distance occurs at noon on the winter solstice. The average of these two extreme angles is similar to the Sun's zenith at the equinoxes, when the Sun is directly overhead at the equator. On these days the Sun's zenith distance measures the latitude, which can also be computed by averaging the extreme zenith angles at the solstices (Figure 2.10, right).

For thousands of years astronomers have observed the Sun, measuring zenith distances and developing accurate declination tables for relating time of year, zenith distance, and latitude. With these tables, a fifteenth-century navigator could measure the Sun's declination, check the calendar, consult his tables, and estimate latitude to within two degrees.[14]

Polaris, a much more distant star than our Sun but clearly visible at night under clear skies, provided another celestial signpost for determining latitude. Called the "Polar Star," Polaris is a bright star that is observed low in the sky near the equator and higher in the sky the farther north one travels. Its angle of elevation above the horizon is the rough equivalent of latitude, and it was used by the sailors of Columbus's era to indicate both latitude and general direction.[15] More accurate estimates of latitude were possible in the sixteenth century with the *nocturnal*, an instrument that corrects the systematic deviation throughout the year of Polaris from a position directly above the North Pole.[16] The navigator would

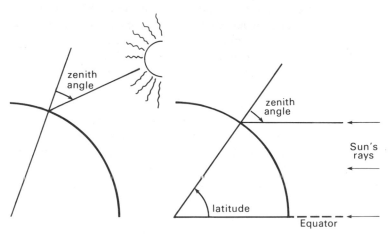

FIG. 2.10.   The zenith angle (left), between the Sun's rays at noon and a straight line from the center of the Earth, is equal to latitude (right).

(1) sight on Polaris; (2) adjust a movable arm to point at Kochab, the other bright star in the constellation Ursa Minor, or Little Bear; (3) note the time of year on the nocturnal's large disc; and (4) read his corrected latitude. More exact corrections were possible in the eighteenth century, when he could tell the time since noon with a ship's chronometer.

Useful in estimating latitude, time was indispensable in determining longitude, which is readily computed from the difference in time between two places. The Earth rotates through 360 degrees in a day and through 15 degrees in an hour. Conversion between a time difference and longitude is simple—provided one knows the precise time both locally and at the prime meridian. If the local time is noon—the Sun is at its highest point in the sky, its zenith for the day—and if the local time at Greenwich is 2 P.M., the place is 30 degrees west of Greenwich. The Sun travels from east to west over the Earth as the Earth rotates into the east from the west (Figure 2.11). If the day were more advanced at

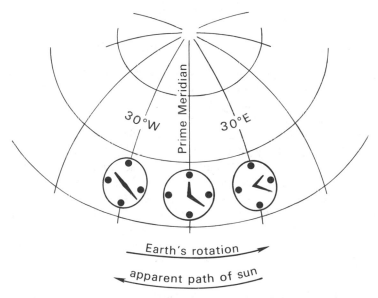

FIG. 2.11.   As the Earth rotates from west to east at 15 degrees per hour, the Sun moves from east to west. When the time is 12:20 P.M. at the prime meridian, local time is 10:20 A.M. at 30 degrees west and 2:20 P.M. at 30 degrees east.

Greenwich, the place in question would be west of the prime meridian. When the day is not as advanced at Greenwich, the place has an east longitude.

Determining longitude from a time difference is not as easy as it might seem, particularly when the navigator is moving, at sea or in the air. To be sure, Greenwich mean local time can be broadcast by radio, as it has been since 1910.[17] Local noon might be judged at a fixed location as the time when the gnomon on a sundial casts the shortest shadow—but only at the equinoxes! Because the equator is tilted 23½ degrees away from the plane of the Earth's elliptical orbit about the Sun, the length of the day varies slightly throughout the year. A clear distinction must be made between mean time, which divides the time for a complete revolution about the Sun into 365¼ days of equal duration, and solar time, which defines the day as the interval between the local noons on two successive days. Solar time varies as much as 16 minutes, 23 seconds per day from mean time. This difference in length between mean and solar days, which is nil only at the equinoxes, is called the Equation of Time.[18]

Satellites, the Doppler Shift, and Instant Location

For thousands of years navigators using celestial navigation have looked to the heavens for accurate estimates of the time difference needed to compute longitude. They have seen not just the Sun, but also the Moon, several nearby planets, and some of these planets' satellites. The pattern of stars varies day by day and, if you observe carefully, by the hour and even by the minute. Observations made over many years permitted accurate predictions of not only sunrise, sunset, and the phases of the Moon, but also of the frequent eclipses of Jupiter's four bright satellites.[19] Almanacs can be used to relate these observations to local time and even to Greenwich time, which in the nineteenth century could also be estimated at sea with an accurate ship's chronometer. The rotation of the Earth through 15 minutes of longitude every minute and through 15 seconds of longitude every second underscores the importance of a highly accurate chronometer or very precise celestial observations.

The nautical almanac is the key to celestial navigation. The navigator first identifies a small number of stars, planets, or satellites with a star chart appropriate to his general latitude, time of year, and time of day. Most almanacs provide data on the Sun, the Moon, four planets, and 50 to 100 stars.[20] Angles are measured and time of observation noted. After additional calculations to correct for atmospheric refraction and other distortions, the navigator uses the almanac's tables to expedite his arithmetic as well as to relate the observed positions of celestial bodies to their "scheduled" positions. The convenience and widespread use by sailors from many nations of the British *Nautical Almanac,* with time published for the Greenwich Observatory, led to the international adoption in 1871 of the Greenwich Prime Meridian as the standard for longitude on sea charts.[21]

A new mode of celestial navigation based on man-made satellites is rapidly replacing the sailor's traditional reliance on heavenly bodies, angular measurements, printed almanacs, and hand or slide rule calculations. This all-weather system, called the Navstar Global Positioning System and developed by the United States Department of Defense, is designed to provide positional calculations accurate to the nearest 2 cm.[22] Great changes are thus in store for the land surveyor as well as for the pilot and navigator. "Revolutionize" is perhaps too modest a term to describe the effect of the Global Positioning System upon celestial navigation: artificial satellites will replace celestial bodies, radio receivers and computers will replace angular measurements and hand calculations, and the satellites will broadcast their own almanacs.

The Global Positioning System is based on a principle in physics called the *Doppler shift,* and the method is described more generally as the Doppler satellite positioning technique.[23] The blasting air horn of a railway engine provides the classic illustration of the Doppler shift: as the train approaches a stationary observer, the sound waves from the air horn appear shorter and more frequent and produce a higher pitch than if the distance between horn and ear were constant. Later, as the train moves away from the observer, the sound waves appear broader and less frequent and produce a deeper tone. Each wave crest is emitted at a single instant in time, and as the transmitter (the horn) ap-

proaches the fixed receiver (the ear), its forward notion reduces
the distance between successive crests and, in effect, produces
shorter waves that are received more frequently than if both transmitter and receiver were not moving (Figure 2.12). Movement of
the transmitter away from the receiver adds to the wavelength and
produces less frequent waves.

The Doppler shift can be observed away from as well as directly
astride the railway. The net effect at an observation point away
from the track is a systematic reduction in the apparent speed of
the locomotive. A critical variable at any instant is the angle A
between lines radiating from the observer to (1) the oncoming
horn and (2) the closest point along the track (Figure 2.13, left).
When angle A is close to 90 degrees, the Doppler effect at the
offset observation point is similar to that at the closest trackside
point. As the train approaches, the shift will become less and less
apparent.

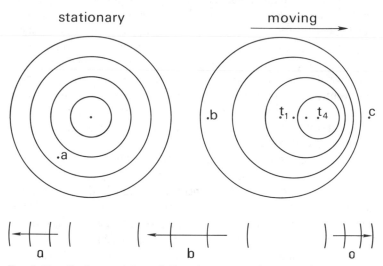

FIG. 2.12.   Stationary air horn (left) emits a concentric pattern of sound waves,
which are received at (a) as a series of evenly spaced wave crests (below, left).
Moving air horn (right) emits an eccentric pattern of circles with centers progressively farther toward the right. Wave crests received at (b) are progressively farther apart (below, center); wave crests received at (c) are progressively closer
together (below, right).

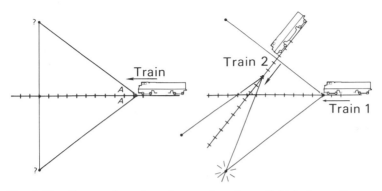

FIG. 2.13. Decreased spacing of waves from horn of single locomotive approaching off-track position can be used to estimate angle A, which yields two possible locations of observing station (left). Similar information from trains on two intersecting tracks can fix the position of the observer (right).

If the observer knows the speed of the train and the original wave frequency of the horn, he can estimate angle A. If he measures the Doppler shift at two different times, he can estimate how far he is from the track. If he knows the precise location of the train at the time of observation, he can narrow down his own location to two points (Figure 2.13, left). Without a more bulky directional microphone or moving his position, he cannot tell whether he is to the left or the right of the train—he knows that he is at one of two points but cannot tell which.

A solution is to measure the Doppler shift from not one train but two, each operating along a different track (Figure 2.13, right). Only one of the two possible locations calculated for the first railway will also be a possible solution for the second. The receiver must, of course, distinguish between the two trains, and the trains must be on schedule. For this system to work within a large region would require a moderately dense railway network with frequent trains, each with a different whistle. For convenience these trains might broadcast codes for their locations, their speeds, and the original frequencies of their air horn. A navigator could thus estimate his location anywhere within the region. To check his estimate and increase precision, a third train, or perhaps even a fourth, would be useful.

The preceding mildly farcical, pedagogic pipe dream—valid in theory but totally impractical because of excessive cost, not to mention noise—would be of little value to marine or air navigation. Yet the underlying theory of the Doppler shift and the geometric relationships between a fixed receiver and multiple moving transmitters is being implemented well above the Earth by a "constellation" of at least 18 orbiting satellites, each broadcasting its own nautical almanac, or *ephemeris*. Accuracy depends upon the precision of the receiver and its antenna, the number of satellites passing nearby, and the number of independent measurements made by the receiver.[24] The navigator of a moving vessel or aircraft may be content with the additional information provided by a single reading, accurate to about 40 m [130 ft], whereas a geodesist needing a location accurate to the nearest 2 cm may need to "occupy his station" for several hours.[25]

Sonar and Loran, Navigation and Mapping

Other on-board electronic equipment used by sailors also aids in the production of marine charts. Sonar, short for SOund Navigation And Ranging, can detect submerged rocks and submarine bars, and can also provide accurate depth soundings for mapping. Loran, an acronym for LOng RAnge Navigation, enables navigators and hydrographers to establish their position by taking "fixes" on widely separated coastal radio transmitters. Both Sonar and Loran use wave geometry to detect relative position.

Sonar is called an *active* sensing system because it receives the reflections of the waves it transmits. Sonar takes "echo soundings" by generating sound waves in water and measuring the time taken by the echos to return to a microphone under the vessel (Figure 2.14). Depth to the sea floor can be computed as half the time interval between transmission and return multiplied by the speed of the waves. Sound waves in water travel at approximately 1500 m/sec [c. 5000 ft/sec], with some variation related to water temperature.

More sophisticated Sonar systems used for navigation radiate their signals horizontally as well as vertically.[26] Visual displays similar to the screen of a radar receiver provide a full 360-degree

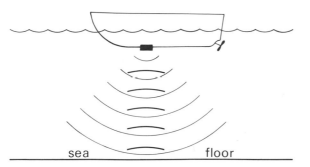

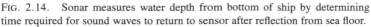

FIG. 2.14.   Sonar measures water depth from bottom of ship by determining time required for sound waves to return to sensor after reflection from sea floor.

sweep around the vessel and indicate distance to the bottom of the channel and other prominent objects such as submarines, anchored mines, schools of fish, and submerged masses of ice.

Radar is a similar type of active sensor that also measures distance by receiving the reflections of its transmissions. Radar waves are shorter, higher frequency electromagnetic radiation, whereas Sonar waves are longer, much lower frequency acoustic waves.

Loran is a *passive* system; an observation station does not generate its own waves but instead receives radio-frequency electromagnetic waves from pairs of coastal transmitters.[27] Each pair consists of a *master station* broadcasting a continual series of pulses at a constant rate and a *slave station* rebroadcasting these pulses at the same rate. The two sets of pulses are not broadcast simultaneously, but with a known, fixed time delay **u**, as shown in Figure 2.15. The concentric circles around master station M show pulses that have been propagating outward for **a, 2a, 3a, 4a, 5a,** and **6a** time units. Slave station S receives these pulses **5a + u** time units after their generation at M. When the slave station rebroadcasts the signal from the master station, the concentric circles around S represent pulses that have been propagating since their original broadcast for **6a + u, 7a + u, 8a + u, 9a + u, 10a + u,** and **11a + u** time units. If each original wave is a distinctly different, coded pulse, it is possible to measure the time lag between the receipt of the original wave from M and the receipt of its rebroadcast version from S. Note that the time differ-

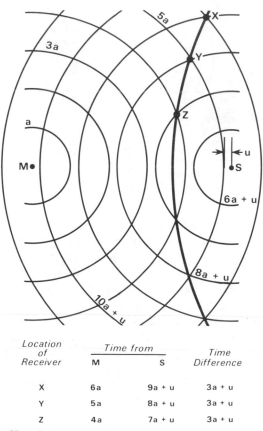

Location of Receiver	Time from		Time Difference
	M	S	
X	6a	9a + u	3a + u
Y	5a	8a + u	3a + u
Z	4a	7a + u	3a + u

FIG. 2.15.   Hyperbola through X, Y, and Z is locus of all points at which time difference is **3a + u** for pulses broadcast from M and rebroadcast **5a + u** time units later from S.

ence is the same for receivers at positions X, Y, and Z, and for all other points along a parabola through these points.

   Pulses from the slave and master stations establish hyperbolic *lines of position,* each with a different, measurable time difference (Figure 2.16, left). All points along the same line of position have the same time difference in the arrival of pulses from the slave and master stations. The navigator's receiver measures a time difference that specifies a particular line of position shown on a Loran

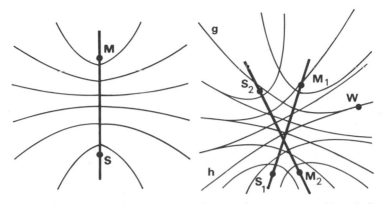

FIG. 2.16.   Set of master and slave stations produces a pattern of hyperbolic lines of position, each representing a unique time difference (left). Time differences **g** and **h**, measured for station pairs 1 and 2, respectively, can be used to fix position of receiving station W at intersection on Loran chart of corresponding lines of position (right).

chart for the two stations. A reading for a second pair of stations provides a second hyperbolic line of position. The intersection of these two lines of position then indicates the location of the plane or vessel (Figure 2.16, right).

Precision receivers permit the accurate estimation of distances and position. Loran can estimate geodetic position, that is, latitude and longitude, to within 900 m [onc-half nautical mile], and can be used to estimate local distances to within 15 to 90 m [50 to 300 ft].[28] Accuracy is greatest along the base line directly between the two stations. The farther from the base line the observer, the greater the position error associated with measuring the time difference between pulses.

Loran can improve the conventional hydrographic chart as well as make it obsolete. By providing surveyors on land or at sea with accurate estimates of position, Loran can be useful in making maps. Loran can also replace maps: some receivers have a digital computer and converter that provides a direct digital readout of location coordinates. Most current users of Loran must still rely upon maps, particularly upon hydrographic charts containing sets of Loran's hyperbolic lines of position. Yet, low cost, highly accurate receiver-computers eventually will eliminate the need for these charts. Navigation maps will still be needed for route planning, but with Sonar and Loran the role of maps in route following will be reduced.

Cruise Missiles: Digital Maps and the Unmanned Bomber

Computers "replace" humans for a variety of reasons: labor cost, accuracy, reliability, speed, and, in some military tasks, danger and possible loss of life. The guidance system of the cruise missile, essentially an unmanned bomber, is computer controlled not only for reliability but also for strategy and tactics—the cruise missile, like the ballistic missile, is intended to penetrate enemy territory and, in military parlance, to "evaporate after delivering its warhead." Although a ballistic missile might be guided only for the first 5 minutes of a 20-minute, supersonic intercontinental trajectory, the cruise missile may be airborne for six hours or more and must be guided throughout its subsonic flight.[29]

The cruise missile is made possible by (1) a digital map of the route to its target, (2) a terrain sensor that compares observed and anticipated terrain features, and (3) an on-board computer that updates the position of the missile in flight and makes any necessary corrections to the course. The terrain sensor is a radar altimeter that scans the terrain below and just ahead of the missile (Figure 2.17). Elevation above sea level is estimated for each cell of a square grid. The computer then compares these sensed elevation data with a stored elevation grid for the immediate and surrounding area. The grid of elevations sensed by the altimeter is positioned like a template over a portion of the digital map. Different alignments and orientations are attempted until a near-perfect match detects the actual course of the missile across the area covered by the stored map. The degree of match is measured by summing the absolute deviations between altimeter readings and corresponding elevations on the terrain map.[30] Rows and columns need not be aligned as the template of sensed elevations is moved over the terrain map in search of a position with an even lower sum of absolute deviations. The computer resets the missile's guidance system to correct any discrepancy between the planned course and the actual course.

Terrain Correlation Process

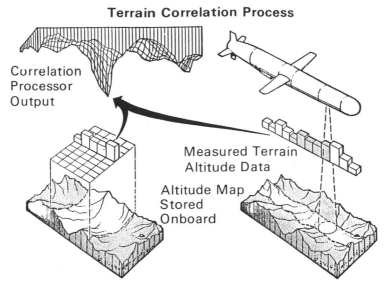

Correlation
Processor
Output

Measured Terrain
Altitude Data

Altitude Map
Stored
Onboard

FIG. 2.17. Radar altimeter of a cruise missile scans ahead to measure distance to ground and generate grid of relative elevation values to compare with stored digital terrain maps.

Source: Courtesy Joint Cruise Missiles Project Office, U.S. Department of Defense.

Two other approaches exist for directing cruise missiles to their targets. One approach is the Global Positioning System described earlier. A constellation of 24 Doppler-shift satellites will place at least four satellites "in sight" at any one time and thus allow instantaneous estimates of position in three dimensions and accurate to within 10 meters. Closely spaced readings can be used to estimate velocity, another parameter used in computing a course correction.

The navigation satellite system is most appropriate for short-range, tactical cruise missiles, which are in flight too briefly for terrain comparison to be useful, or for missions over terrain too flat for reliable terrain comparison. Although radar homing alone is fully adequate for guiding a missile to a very close target, satellite positioning is a more suitable means of directing a missile to a known target beyond the line of sight—radar cannot "lock in" on objects below the horizon. Satellite positioning is also the only practical system for guiding a cruise missile great distances over water to a coastal target—terrain comparison is of no use at sea, where there is no terrain.

The other approach uses microwave or infrared sensing systems able to detect man-made landmarks such as structures, roads, and railways as well as lakes and rivers.[31] Altimeter-based terrain comparison is thwarted by flat and even low-relief terrain. Computerized correlation techniques that relate the microwave reflectivity of ground features to stored digital representations of elevation, culture, and hydrography will greatly extend the range of the cruise missile in areas sufficiently varied to provide useful landscape "signposts."

Several versions of the cruise missile are contemplated, with the principal distinction among (1) air-launched missiles fired from, say, a B-52 or B-1 jet bomber, (2) sea-launched missiles fired from a submarine, and (3) ground-launched missiles fired from a truck-mounted launcher for a short-range, tactical strike. All but the air-launched missile would have a rocket booster. Flight altitude would vary depending upon distance between target and launch point, anticipated atmospheric turbulence and terrain roughness, and the need to avoid surface-to-air missiles and enemy fighters. Because it is much slower than a ballistic missile, the cruise missile often flies at low altitudes to avoid detection by radar. Digital vertical obstruction data are used to avoid smoke stacks, transmission lines, and other objects projecting well above the land surface.

A typical mission might begin at sea with a long-range missile launched from a torpedo tube of a submerged submarine. Booster rockets lift the missile to 3 km [2 mi], the cruising altitude for the first 1,500 km [900 mi] of the mission. Its diameter of only 53 cm [21 in.], length of 6.24 m [20 ft], and volume of 1.37 cubic m [48 cubic ft] provide a radar cross section far smaller than that of a manned bomber.[32] As the missile approaches the coast, it descends to an altitude of 20 m [70 ft] to escape radar at landfall. Its altitude inland is about 50 m [160 ft] in hilly terrain and 100 m [330 ft] in mountainous regions (Figure 2.18).

The route is planned in advance, and the missile is provided with perhaps 20 digital maps.[33] Grid cells may be as small as 10 m [30 ft] on a side, and elevations are accurate to the nearest

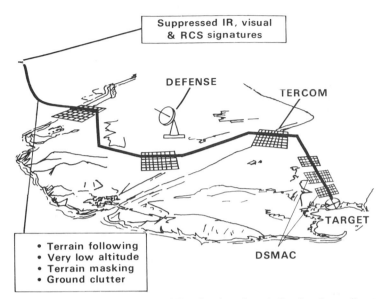

Suppressed IR, visual & RCS signatures

DEFENSE

TERCOM

TARGET

- Terrain following
- Very low altitude
- Terrain masking
- Ground clutter

DSMAC

FIG. 2.18.   Generalized profile of Tomahawk cruise missile after descending to low altitude as it approaches coast. Average altitude is 3 km [1.9 mi] over ocean, 20 m [66 ft] at landfall, 50 m [160 ft] in hilly areas, and 100 m [330 ft] in mountainous terrain. TERCOM (Terrain Comparison) and DSMAC (Digital Scene Matching Area Correlator) systems guide missile through electronic correlation of ground conditions with stored digital maps of terrain and surface features.

Source: Courtesy Joint Cruise Missiles Project Office, U.S. Department of Defense.

3 m [10 ft] or less. The first map covers a belt 10 km [6 mi] wide, broad enough to permit a match despite whatever drift accumulated during the long flight over water. Cruise missiles can drift off course because of stormy weather, locally strong crosswinds, and minor engine malfunctions. Maps used for subsequent periodic checks of the missile's location cover belts narrower but longer, perhaps 2 km by 10 km [1.2 mi by 6 mi].

Terrain comparisons are needed several times during the missile's 1,200-km [750-mi] overland flight. Course corrections are attempted where the programmed trajectory changes its direction and after intervals over which appreciable drift might have accumulated. Without digital maps and the altimeter to direct the autopilot, the cruise missile might be tens of kilometers off target.[34] With maps stored and read by its on-board computer, the missile can deliver its warhead to well within 100 m [330 ft], about half a city block from its target (Figure 2.19). With accuracy this high, the warhead of a cruise missile need not be as powerful and lethal as that of a ballistic missile.

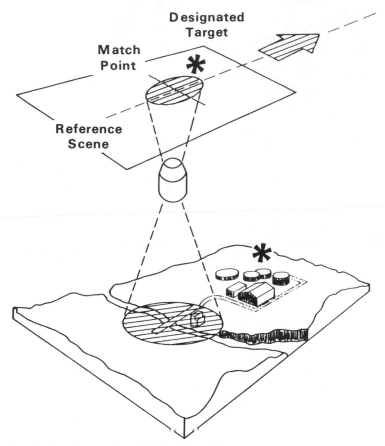

FIG. 2.19.   Cruise missile's Digital Scene Matching Area Correlator (DSMAC) detects designated target by matching sensed ground features with digital representation carried onboard.

Source: Courtesy Joint Cruise Missiles Project Office, U.S. Department of Defense.

Map Use, Human Factors, and Optimal Displays

Not all map reading tasks are as rote and easily programmed as the terrain comparisons of the cruise missile. The pilot of a jet fighter, manned bomber, or reconnaissance aircraft might have a mission without a set, predefined course. Flexibility requires the display of geographic information in the cockpit when needed. Success of the mission requires that the display be as complete and as readily comprehended as possible. Effective cartography thus depends upon *ergonomics,* also called *human factors engineering,* a cross disciplinary mix of applied psychology and operations research.

Poor lighting in the cockpit warrants special maps for air navigation. Illumination is a particular problem for night flying.[35] Lighting the map with standard white light can interfere with the pilot's noting important landmarks below. One solution, proposed as early as 1923, is red lighting in the cabin and red coding on the map of features not needed for night flying. Less intense red cabin lighting is unlikely to interfere with the pilot's ability to detect prominent white-light ground features. Mapped features printed in red are invisible at night under red light in the same way that features printed in white ink on white paper would be invisible during the day under white light. The pilot can thus focus upon only those symbols useful at night.

Other solutions exist, and there is considerable need for a careful evaluation of in-flight conditions and the limitations of the human eye-brain system.[36] Some map features, for instance, might be printed with a self-luminous fluorescent ink, visible under ultraviolet cabin lighting. Although ultra-violet lighting produces glare and visible reflection from interior surfaces, this ink can glow in the dark for up to an hour after the light is turned off. Hence the light source would, for the most part, not interfere with external cockpit vision. Unfortunately, the brightness of printed fluorescent lines is not uniform, and crews have sometimes experienced severe visual fatigue. Other, more promising solutions include self-illuminated cathode ray tube (CRT) displays and back-lighted displays projected from a roll of film.

All-electronic systems, based on the same principles as color television, are the most promising. Without moving parts, they are more responsive and reliable than optical-mechanical systems. Scale can be changed rapidly, with features selected and deleted according to need, and brightness and hue adjusted for day or night use. Unlike filmstrip and microfiche systems, electronic systems can readily accept and display new data, perhaps from onboard radar and other sensors. Similarly, obsolete data—a hazard in combat—can be deleted instantaneously.

Electronic displays pose new problems for the cartographer and the human factors engineer. Which colors provide the maximum contrast? Which features merit maximum contrast? To what extent might switching colors produce confusion? How important is it that conventional assignments of colors to features be maintained on CRT displays? How effectively can flashing symbols emphasize highly important features? What can be done to minimize fatigue? What are the limits of vigilance and perception? How much accuracy and reliability might be expected from man, machine, and map?

Detail and information content is another important factor in designing navigation maps. Aeronautical charts are particularly sensitive to flight speed and whether ground features or instruments are to guide the aircraft along its course. Sectional Charts for low-to-medium speeds might be printed or displayed at 1:500,000 and contain more local details than, say, Jet Navigation Charts at 1:2,000,000. Charts intended for flying according to visual flight rules (VFR) emphasize landmarks, terrain, and airport and airspace information. Charts designed in accord with instrument flight rules (IFR) emphasize the radio navigation system and instrument approach systems around airports. A variety of charts are in use for various air speeds, day or night flying, VFR or IFR navigation, and route planning or route following.[37] Electronic retrieval and display systems for navigation can avoid the inflexibility of the printed map and provide pilots with a ready library of easily handled, truly relevant map information.

"Trickle Down"

Navigation and military applications will continue to exploit computers and other electronic systems to improve some maps and replace others. As in many areas, the perceived need for ever better national security will remain the principal impetus for major new developments in mapping and map use. A good defense system requires accurate geographic intelligence and accurate navigation. Only the comparatively lavish appropriations for defense are likely to sustain or increase the current rate of development in digital cartography. Mapping thrives on war and threats of war.[38]

Technology perfected for the military is often adapted for civilian use, and digital cartography will become ever more common in other branches of government and in the private sector as well. The Global Positioning System and Loran will eventually benefit the surveyor and the weekend sailor.[39] Digital terrain models will benefit the engineer, the petroleum geologist, and the regional planner. Human factors research on the effectiveness of cartographic displays, both video and paper, will "trickle down" to commercial and private aviation, to the users of geographic information systems (discussed in Chapter 5), and even to tourists, users of atlases, and students.

Map use and the form of the map will continue to evolve. This evolution may be more rapid than might now seem possible. In 1950, for example, the agricultural scientist never considered the possibility of photographs from man-made moons; in 1980, he was estimating acreages from Landsat imagery. In 1960, the land surveyor was delighted with the expensive, heavy electric desk calculator; in 1980, he could own several electronic calculators and even carry one around in his wallet. Before 2000, he might well push a button on a small, portable pocket navigator and read out his coordinates accurate to several decimal places. Knowing where you are and how to get to where you want to go will no longer be challenging problems.

CHAPTER 3

Boundaries and Surveys

Geographic position, important to the traveler and invading army alike, is doubly significant to an economy that recognizes land tenure, whether vested in private citizen or people's collective. The value of a parcel of land depends upon, among other things, an accurate description of its boundaries: an owner needs to know the extent of the land he is buying and where he may erect a fence. Good boundaries not only make good neighbors but also help prevent disputes ranging from bitterness among homeowners arguing about a hedge to war among countries contesting an international boundary. In addition to promoting the sale and control of land, boundary delineation and land recording systems provide a basis for real property taxation and thus permit governments to collect monies in support of education, defense, public works, and relief for the poor. Surveying as a profession is nearly as old as cities, governments, and other institutions that depend upon land records.

This chapter examines the evolution of land measurement technology from the primitive ancient Egyptian devices for surveying boundaries through modern electronic systems for measuring distances with light beams, estimating locations in a control survey network with statistical algorithms, removing geometric distortion from air photos with computer-controlled photo printers, and surveying with inertial guidance systems. Surveys are needed both to describe boundaries for deeds and treaties and to place on the landscape such visible markers as monuments, fences, walls, and roads. Modern technology is helping the surveyor in the office as well as in the field—in calculating areas and producing maps as well as in measuring distances and angles. This chapter focuses on the latter activity, the acquisition of basic data about the horizontal and vertical positions of Earth features. The land inventories and the display and analysis systems discussed in later chapters depend upon field surveys, as do intelligence satellites and Doppler positioning systems. Geographic data are most useful when tied to the land by accurate coordinates that permit the systematic integration of information collected by different organizations for various purposes in widely scattered locations.

Angles and Distances

Brief accounts for the history of surveying invariably focus upon the evolution of instruments for measuring distances and angles and for sighting along an accurate horizontal line.[1] The practice of surveying dates at least to 3000 B.C., although few descriptions of instruments and practices have survived. The ancient Egyptians, and probably the Babylonians as well, had highly accurate surveys based in part on cords with evenly spaced knots. Retracement surveys were needed to reestablish field boundaries after the annual flooding of the Nile Valley, and the pyramids were characteristically built upon a nearly perfect square base oriented to the true cardinal directions.[2] The Greeks advanced the art of land survey still further, although their geometry, literally "earth measurement," was more abstract than practical. Their *diopter,* a sighting device with a graduated disc for measuring angles, was useful for taking altitudes and for leveling, the measurement of vertical distances between points.

The Romans, aggressive in the construction of roads and aqueducts as well as in military conquest, were skilled engineers and surveyors. They developed several highly useful instruments. The *groma,* two perpendicular horizontal rods with plumb lines descending from each end, was used in the field to lay out square corners. The *chorobates,* a straightedge about 6 m [20 ft] long with a groove filled with water on top, permitted accurate sighting along a horizontal line. A surveyor could sight along this level and direct his coworkers in positioning a plumb bob farther down the line. The *libella,* an A-frame with a plumb bob suspended from its apex, also served as a level. Maps usually represent horizontal distances measured with leveling instruments based on these principles.

Both the Greeks and Romans wrote lengthy and thorough treatises on surveying, most notably *The Dioptra,* by the Greek Heron (c. 120 B.C.), and the *Codex Acerianus* (c. 550 A.D.), containing excerpts from a still earlier manuscript by the Roman, Sextus Julius Frontinus (c. 40–103 A.D.). As with most Greek and Roman contributions to science and mathematics, the principles of surveying survived the Middle Ages in the custody of the Arabs. Further progress in instrumentation was marked by the invention in the early seventeenth century of the telescope, which increased both the speed and precision of surveys; the *Gunter chain,* which included 100 links in its 66-foot [20.1-m] length to expedite measurement along boundaries; and the vernier, which increased the precision with which a graduated scale for measuring angles could be subdivided.[3]

Like many advances in mapping in general, advances in surveying instruments and practices coincided with advances in the art of warfare, especially with improvements in gunnery in the sixteenth century. Prior to gunpowder and the cannonball, military assaults depended largely on powerful crossbows and giant slingshots for heaving rocks at a relatively secure enemy behind thick castle walls. The advent of cannon that could loft cannonballs over high curtainwalls or chip away protective masonry led for awhile to higher castles, with thicker walls as well as shapes that permitted a more thorough coverage of the surrounding area with crossbow and defensive cannon.[4]

A gunnery officer was partly a surveyor: in addition to determining the appropriate amount of gunpowder, he had to estimate distance to his target and calculate an appropriate elevation angle for his cannon. Seige warfare was slow, and he could usually afford the delay of trial-and-error "zeroing in" on the target with a series of successive shots separated by sufficient time to make a minor adjustment in charge or elevation. The *gunner's quadrant,* invented in 1537 by the Italian Nicola Fontana, was intended to simplify these calculations. The gunner would insert one leg of a triangle into the bore of the gun and read the elevation angle of the cannon from the position of a plumb line against a curved scale (Figure 3.1, left). By adding or removing wedges beneath the barrel, he could increase or decrease the elevation angle.[5]

Another helpful measuring instrument was the *cross-staff*, or *baculum*, a graduated scale with a sliding perpendicular cross piece of fixed length. After positioning the cross piece to intersect lines of sight to both the top and base of a distant vertical object, the gunner could then read the object's elevation from the graduated scale (Figure 3.1, right). This instrument was introduced to Europe from earlier Arab writings translated in the fourteenth century by Levi ben Gerson.[6] Numbers marked on the scale were related to the tangent of half the angle between the two lines of sight. A single reading was worthless unless the distance from the object was known, yet two readings, taken a known distance from each other, could be used to calculate not only vertical elevation but also horizontal distance.[7]

Geometric calculations can be digital or analog—arithmetic or graphic. For many applications, including some in surveying and gunnery, distances measured with difficulty in the field can be determined safely if not with high accuracy from a scaled-down drawing carefully plotted from a few precisely measured angles

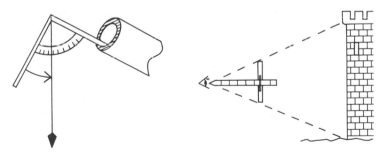

FIG. 3.1.   Gunner's quadrant (left) was inserted in the bore of a cannon so that the angle of elevation could be read against the plumb line. Cross-staff (right) used graduated scale and movable crosspiece to estimate elevations of distant objects.

and distances. With the aid of an *open-sight alidade* and *plane table,* developed during the sixteenth century, a surveyor could plot horizontal angles directly, without measuring them.[8] Consider, for example, the problem of mapping three points, only two of which are easily accessible or safe, but all of which are intervisible. Assume that the horizontal straight-line distance between the two accessible points, A and B, is known or can be measured. This "base" distance A–B is plotted first, at the intended scale of the map, as a–b. The plane table, mounted on a tripod so that its surface can easily be leveled, is positioned successively at points A and B (Figure 3.2). The surveyor orients the table horizontally by aligning the line a–b on the drawing with a line of sight from A to B. He then uses the alidade to find and plot the line of sight from A toward C. Using a similar set of observations at point B, from B back to A to orient the plane table, and from B toward C to plot the angle ABC, the surveyor then finds the horizontal position of C at the intersection of the scale-model lines of sight

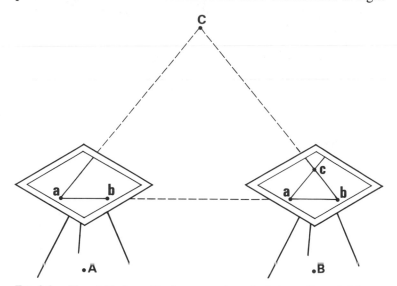

FIG. 3.2.   Plane table is used to draw accurate angles between lines of sight, so that distant points can be plotted by intersection. Map represents lines of sight from A to B and C and from B to A and C; lines a–c and b–c intersect at the to-scale position of C.

from a and b. This process, called *intersection,* can be repeated
to include additional points. Distances such as A–C and B–C can
be determined from the scale of the drawing and the lengths of
their corresponding lines, a–c and b–c.

Although the *relative* accuracy, or shape, of the plane table map
depends upon the angular measurements, the *absolute* accuracy,
shape *and* size combined, depends as well upon the accuracy of
the measured horizontal distance from A to B. The surveyor must
be careful to measure ground distance along a level line; otherwise, the horizontal distance of the base—and the scale of the
entire drawing—will be inaccurate. When there is a substantial
difference in elevation between these two points, it is necessary to
divide the traverse into short, manageable segments (Figure 3.3).

With a telescopic sight and *level rod,* a graduated stick held
vertically by an assistant, the surveyor could also read the elevation difference for each segment if the graduations on the rod were
adjusted for the height of the sight. These elevation differences,
accumulated along a lengthy traverse, are used to calculate relative elevations for prominent features.

Arithmetic approaches to triangulation proved more accurate
than graphic construction. Although a few earlier writers have
discussed the applications of trigonometry to surveying, Wille-

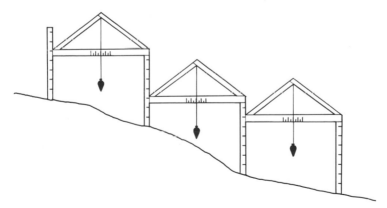

FIG. 3.3.   Graduated vertical rod and libella illustrate principles of leveling, the
measurement of both horizontal and vertical distances between points.

brod Snell (1591–1626), a Dutch mathematician, developed the first practical mathematical approach to triangulation.[9] In a survey conducted in 1615, Snell used the angles and length of one side of a triangle to calculate the lengths of the other two sides. He also used triangulation to estimate the length of an arc along the Earth's meridian.

Trigonometric tables, logarithms, and precision field instruments for measuring angles assured the widespread adoption of mathematical triangulation by surveyors. Precise tables of such trigonometric functions as the *sine,* the ratio of the length of the side opposite a given accute angle in a right triangle to the length of the hypotenuse, had been printed as early as 1533.[10] The invention of logarithms in the early seventeenth century by John Napier (1550–1617) greatly simplified the numerous calculations needed to find the unknown sides of oblique triangles.[11] Accurate measurement of angular differences between lines of sight was possible, after 1550, with the invention by an English surveyor, Leonard Digges, of the *theodolite,* an instrument with a sighting device and graduated circles for measuring angles in both the horizontal and vertical planes. The *transit,* a simpler version of the theodolite, usually including a magnetic compass but designed originally to measure only horizontal angles, was invented about 1830.

Lasers and Distance Measurement

Surveying has made considerable progress in measuring distance since the ancient Egyptians and Babylonians discovered the convenience of cords with uniformly spaced knots. The portability and durability of the Gunter chain, with its 66-foot [20.1 m] length divided into 0.66-foot long links, served the surveyor quite well since its invention in 1620. A rectangular plot ten chains long by one chain wide contains 43,560 square feet [4047.0 m²] or exactly one acre [0.4 ha]. It also left an imprint on the landscape: many older highway rights-of-way are one, two, or three chains wide. A more recent development is the *surveyor's tape,* a thin ribbon of steel 100 feet [30.5 m] long and graduated to the nearest 0.01 foot [3 mm]. By the end of the nineteenth century, the steel tape, readily wound onto a reel for easy handling, had replaced the comparatively bulky Gunter chain. With the aid of tables and

calculations to correct for thermal expansion and sag, a steel tape
in the hands of a competent surveying party can measure distances
accurate to within one part in 5,000, or better.[12]

Various triangulation techniques have also been employed in
measuring shorter distances. The *subtense bar,* with a known length,
commonly 2 m [6.6 ft], is mounted horizontally about 1.5 m
[4.9 ft] above the ground on a tripod. A surveyor, up to about
200 m [650 ft] away, aims his theodolite successively at each end
of the bar, records these bearings, subtracts to find the angle "subtended" by the bar, and calculates horizontal distance by multiplying half the length of the bar by the trigonometric cotangent of
half this angle.[13] For greater distances, a "trig traverse" applies
the same principles with a carefully measured base line at a right
angle to the line of sight. The accuracy of the estimate depends
upon the length of the base line and its distance from the theodolite, which jointly influence the precision of the subtended angle.

A transit telescope with two horizontal *stadia hairs* etched into
a glass disc within the telescope is another application of the same
concept. The transit is focused on a graduated, vertically held
stadia rod. The stadia hairs subtend a constant, known angle so
that the length of the distant base of the triangle can be estimated
by the number of gradations on the rod seen between the stadia
hairs (Figure 3.4). Stadia measurements, useful in making topo-

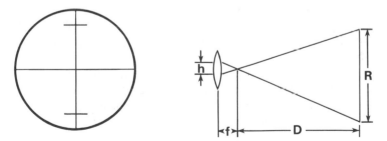

Fig. 3.4.   Upper and lower stadia hairs (left) etched into the objective lens of a
transit telescope are separated by vertical distance **h**. Length **R** of a vertical stadia
rod will be included between the stadia hairs (right). Distance **D** to the rod can
be computed from **h**, **R**, and the focal length **f** as $D = fR/h$. The surveyor reads
R using gradations painted on the stadia rod. The focal length and separation of
the stadia hairs are sometimes designed to yield the simple formula $D = 100R$.

graphic maps, may be as accurate as one part in 1,000, although normal accuracy is only 1 in 500.[14]

Distance is time, or more precisely, a function of the time taken for light or sound to travel the distance to a reflector and return to its origin. As described in the previous chapter, this principle is the physical basis for Sonar, which measures depth to the sea floor by the travel time in water of sonic, or pressure, waves. On land, of course, use of sound waves is impractical: focusing is difficult and the noise would be objectionable. Electromagnetic radiation—light, radio, or microwave radiation, for example—provides an alternative, but with a velocity in the atmosphere 900 thousand times that of sound, its travel time must be measured quickly and precisely. Over a distance of 10 km [6.2 miles], say, an electromagnetic wave completes a return trip in 67 microseconds [67 millionths of a second]; for a distance of only 100 m [328 ft], the corresponding travel time is an even more elusive 0.7 microseconds. For precision to the nearest meter, the round-trip time must be accurate to the nearest 0.007 microsecond.

Surveyors' first use of artificial light was merely to provide triangulation surveys with a target visible at night over great distances. A significant advance was the development, in 1825, of an intense beam of white light by reflecting forward with a parabolic mirror the incandescent radiation from a small fragment of limestone heated by an alcohol flame enriched by a jet of oxygen. Thomas Drummond, an engineer with Britain's Ordnance Survey, developed this light to overcome poor visibility between triangulation stations on two summits in Ireland 106 km [66 mi] apart.[15] Further modifications of Drummond's innovation led to the limelights of nineteenth century theaters.

Little progress was made in the use of artificial light for surveying until the 1940s, when Erik Bergstrand, a Swedish physicist, designed an experiment to measure electronically the speed of light. Bergstrand's approach, an application of *phase shift comparison,* transmits a frequency-modulated beam of light to a set of prisms, which in turn reflects the signal back to a receiver (Figure 3.5). The light is pulsed very rapidly by a pattern oscillator. A photocell detects the returning pulsed beam and converts it to an electronic signal, which is then amplified for comparison

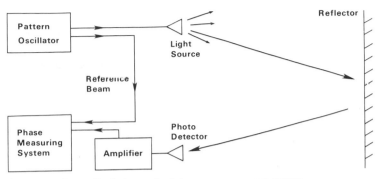

FIG. 3.5. Generalized electronic distance measurement system.

with a reference signal derived from the same pattern as the transmitted beam. The reference and returning beams will be generally out of phase—that is, their respective peaks and valleys will arrive at the phase measuring unit at different times—because the latter does not in general travel an integral number of wavelengths to the object. The two waves can be delayed, one relative to the other; the amount of this delay is a measure of the fractional wavelength to the object. Two or more complementary "pattern frequencies" are used, for instance, a fine pattern of 15 MHz and a coarser pattern of 150 kHz, so that measurement of the phase difference can detect distances between measuring unit and reflector as long as several kilometers with a precision of several millimeters.[16]

In the 1950s, several firms began to market electronic distance measuring (EDM) equipment based on this principle. At first visible light was used, but the improved *geodimeter* employs infrared light, with wavelengths slightly longer than visible light, to penetrate haze. The *tellurometer* uses significantly longer, microwave radiation to reduce further the distortion resulting from moisture in the air.

Early applications of EDM techniques focused upon the control surveyor's need to measure distances much longer than the comparatively short traverses of the land surveyor, and with much greater precision. In 1959, for instance, the Air Force called upon the U.S. Coast and Geodetic Survey, now the National Geodetic

Survey, to assist in the accurate positioning of missile-tracking equipment at Cape Canaveral. Standards called for measurements accurate to within one part in 400,000—four times more stringent a standard than for the most precise, first-order geodetic surveys. Yet the geodimeter not only met but exceeded this standard: probable error was judged statistically to be less than one part per million.[17]

As with many developments of so-called high technology, EDM equipment shortly was made available to a large, receptive market, at a lower cost and with significant improvements. Laser light, which is used now for the geodimeter, provides accurate results in daylight. An automatic level measures the angular deviation from the horizontal of the laser beam, and an internal computer determines both horizontal and vertical distances. In 1981, a surveyor could purchase for less than $15,000 an EDM unit able to measure a distance of, say, 100 m [330 ft] to within 6 mm [0.25 in.]. At only 15 kg [32 lbs], the entire unit can be carried easily by one man.[18]

Surveys once depended almost exclusively upon triangulation for estimating distances that could not be measured directly. Because of their greater accuracy, distance measurements have replaced many angular measurements, and *trilateration*, based exclusively on horizontal distances, is displacing triangulation as the foundation for control surveys covering large areas. Trilateration also uses labor more efficiently: in one comparative study a small survey with a three-person field party required six person-days for angle measurement but only one and a half person-days for the distance measurements needed to compute the same results.[19]

At each station distances are measured to a number of other, visible stations, some of which have "fixed" positions. These fixed stations have known plane coordinates, based upon an earlier, higher-order control survey. The set of measured distances is then used to estimate coordinates for the other observation points by a statistical process called *least-squares adjustment*.[20] The results are obtained as plane coordinates—for which land surveyors once had little use.

The least-squares solution is based upon a system of equations. Each station that is not fixed represents two unknown values, its

X and Y coordinates. Each distance measurement is associated with a linear-distance equation of the form

$$[\text{distance}(a,b)]^2 = [X(a) - X(b)]^2 + [Y(a) - Y(b)]^2,$$

which expresses the square of the distance between points a and b as the sum of the squares of the respective differences in their X and Y coordinates. Each equation has either two or four unknowns—two if between a fixed and an unfixed station, and four if between a pair of unfixed stations. The number of equations minus the number of unknown coordinates yields the number of *degrees of freedom*, which represents the number of "surplus" measurements. These extra measurements serve as useful checks against inaccurate field measurements, random operator error, and instrument errors. All measurements, after all, are only estimates, and confidence in a solution depends upon these independent checks.

Least-squares adjustment is a mathematical method for finding a "best-fit" solution to the set of linear-distance equations. Because of measurement error, when there are more equations than unknowns, no single set of plausible values for the unknown co-ordinates is likely to yield trial point locations with separations exactly matching the measured distances. Nonetheless, some sets of solutions might be considered better than others. Least-squares adjustment defines the goodness of a solution as the extent to which the sum of the squared differences between the measured distances and the computed (adjusted) distances for corresponding pairs of points in the solution is as small as possible—hence the term "least squares." Differential calculus, formalized in the seventeenth century by Newton and Leibniz, was applied in the early nineteenth century by Legendre and Gauss to provide a systematic means for distributing measurement error throughout a control network.[21] Many individual calculations are required, and computational complexity increases as the numbers of distances and unfixed stations increase. Without modern digital computers, the widespread use of least-squares adjustment that makes trilateration practicable would be difficult if not impossible.

Based upon statistical theory, least-squares adjustment provides the surveyor with several useful statistical indexes of accuracy.

The *standard error,* a statistical estimate of the average precision of the adjusted positions of the control points, assesses the overall accuracy of the solution. Individual stations can be evaluated with *error ellipses* that express confidence in a station's adjusted position as a probability.[22] Error ellipses show, for example, the zone within which we might be 95 percent confident of finding the station's true position. Because EDM instruments are highly accurate and the ellipses thus quite small, error ellipses are usually plotted with a grossly exaggerated scale to indicate relative accuracy. Figure 3.6 illustrates the superior accuracy of the survey on the right, with 10 degrees of freedom, as compared to the survey on the left, with only 3 degrees of freedom. As indicated by the larger ellipses, confidence is lower in the adjusted positions of stations from which relatively few distances were measured.

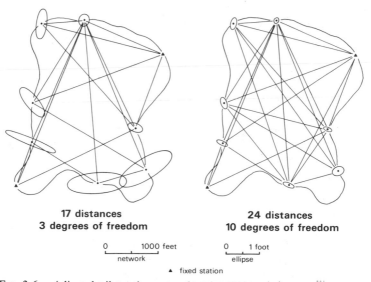

17 distances   **24 distances**
3 degrees of freedom   **10 degrees of freedom**

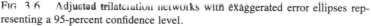

0   1000 feet   0   1 foot
         network   ellipse

▲ fixed station

Fig. 3.6   Adjusted trilateration networks with exaggerated error ellipses representing a 95-percent confidence level.

Source: Adapted from Paul R. Wolf and Steven D. Johnson, "Trilateration with Short Run EDM Equipment and Comparison with Triangulation," *Surveying and Mapping* 34 (1974): 337–46. Reproduced with the permission of the authors and the American Congress on Surveying and Mapping.

Control Networks: Our Locational Infrastructure

Plane coordinates estimated by distance measurement and least-squares adjustment can be no more accurate than the known co-ordinates of the "fixed" stations used for trilateration. The surveyor needs accurate existing stations to which to tie his local, special-purpose surveys, and the establishment and maintenance of a geographically extensive network of control stations conforming to accepted, published national standards of accuracy is clearly a necessary function of government, to be supported collectively for the common good. Control surveys also provide a foundation for systematic, detailed, large-scale mapping, which in turn serves as a base for intermediate-scale thematic maps and geographic information systems.

Until quite recently, control surveys have been based on triangulation, the trigonometric principles of which were devised by the ancient Greeks. The first control survey of a large area was the triangulation survey of France conducted in the eighteenth century by Jacques Cassini (1678–1756) and his son. The triangulation survey was sufficiently complete by 1750 for the Cassinis to turn their attention toward filling in the country's topographic details on a series of 173 sheets covering the country at a scale of 1:86,400.[23]

As with most extensive triangulation surveys, the Cassinis crossed the country with chains, or *arcs,* of triangles organized as quadrilaterals so that sightings could be made at each station, visibility permitting, to five other stations (Figure 3.7). These geodetic arcs

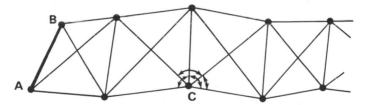

FIG. 3.7. Example of the arrangement of horizontal control stations as a chain of quadrilaterals in an arc of a geodetic triangulation network. Length of base line measured between A and B provides the network with a scale upon which estimated distances between other pairs of stations can be based. Small circular arcs show some of the angles that might be measured at station C.

were routed along coastal plains, where specially constructed towers made distant stations intervisible at night, and through more rugged areas, where peaks were intervisible. Latitude and longitude were checked occasionally by astronomical observation, but the field party mostly measured angles between stations. It was much easier to measure angles than distances, but a few base lines were needed to establish a scale for the network. Base-line distance was measured carefully along level lines, and these measurements repeated several times to confirm their validity.

Control surveys use a hierarchy of networks so that comparatively few highly accurate "first-order" arcs might serve a large area (Figure 3.8). Second-order arcs, fully adequate for most precision surveying but less accurate and less costly than components of the first-order network, fill in gaps between first-order arcs and thereby "densify" the control network. Third- and fourth-order arcs, demanding less accuracy, provide an even denser control network. Triangulation arcs are also supplemented by traverses, series of straight-line segments measured with theodolites and EDM equipment. Separate networks are maintained for *horizontal control,* which indicates geographic position by latitude and lon-

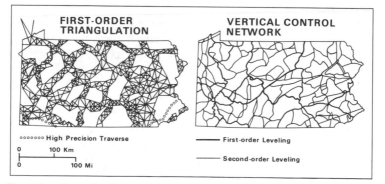

FIG. 3.8.   Generalized maps of geodetic control networks in Pennsylvania, 1980. First-order triangulation and high-precision traverse (left) are supplemented by second-order triangulation, first- and second-order traverses, and second- and third-order control (not shown). First- and second-order leveling lines (right) are supplemented by additional second-order leveling in some localities (not shown).

Source: Status maps maintained by the National Geodetic Survey.

gitude as well as by plane coordinates, and *vertical control*, which specifies elevation above a given datum. In the United States the vertical datum is usually mean sea level.

Each level in this control survey hierarchy has a specific accuracy standard. In a first-order horizontal control survey, for example, distance error may not exceed 1 part in 100,000. This standard should guarantee that a distance computed using latitude and longitude as 1,000 km [1,609 mi] would be accurate to at least 10 m [33 ft]. Latitude and longitude are reported in degrees, minutes, and seconds, with five decimal places for seconds provided for computational purposes only. For second-order surveys, the tolerance is 1 part in 50,000. Standards for third- and fourth-order surveys, not a part of the geodetic network but used in topographic mapping and construction surveys, call for error not to exceed 1 part in 20,000 and 10,000, respectively. Property surveys typically satisfy the fourth-order standard.

Base mapping requires a relatively simple mathematical description of the Earth that can provide a reference surface from which to project horizontal positions onto a *developable*, or flattenable, surface. The Earth is not a perfect sphere, and use of a spherical model is inadequate for geodetic surveying and large-scale mapping. Although suitable for small-scale world and national maps that cannot possibly portray precise horizontal position, a sphere would distort estimates of ground distance between widely separated places as well as yield latitudes noticeably inconsistent with astronomical observations. Yet a workable compromise is needed between the oversimplification of the sphere and the mathematical unwieldiness of the *geoid*, the theoretical, asymmetric surface that would result from extending mean sea level beneath the continents. The *ellipsoid*, also called a *spheroid*, provides a suitable approximation in a three-dimensional figure formed by rotating an ellipse about the Earth's axis of rotation (Figure 3.9, left). In cross section the Earth is like an ellipse: its diameter across the equator, 12,756 km [7,926 mi], is longer than its diameter between the poles, 12,713 km [7,900 mi]. This flattening at the poles is far more prominent a departure from a true sphere than the comparatively modest, pear-like bulge in the midlatitudes of the southern hemisphere (Figure 3.9, right). Ellipsoids

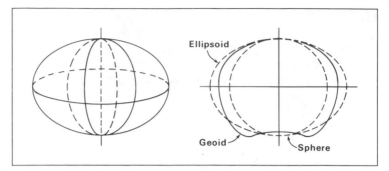

FIG. 3.9.   Ellipsoid (left) is formed by the revolution about the Earth's axis of
an ellipse. The geoid (right) is much closer in cross section to an ellipse than to
a sphere.

can be derived that are separated nowhere by more than about
80 m [270 ft] from the geoid, whereas for a sphere this deviation
would exceed 7 km [4 mi].[24]

A worldwide reference ellipsoid cannot provide for an individual continent the accuracy possible with an ellipsoid chosen spe-
cifically to accommodate regional idiosyncrasies in the Earth's
shape. Consequently, unique ellipsoids were developed in the
nineteenth century for North America and Europe. In North America
geodetic surveying and topographic mapping have been based
upon the ellipsoid developed in 1866 by Alexander Clarke, a
British geodesist. Clarke's ellipsoid has an equatorial diameter
of 12,756,412.8 m and a diameter between the poles of
12,713,167.6 m. This geometric Figure served in turn as a basis
for the *1927 North American Datum,* which is said to tie the first-
order triangulation network to the Earth at a control station in
north central Kansas, on Meades Ranch. This datum assigns each
control station a latitude and longitude estimated by a tedious,
painstaking least-squares "adjustment" similar to the process described earlier. Horizontal distance in this case is computed on the
surface of an ellipsoid. Estimated latitudes and longitudes must
yield computed distances between stations that match field measurements as well as accommodate astronomical observations taken
at selected control stations.

Like most measurement systems over 50 years old, the 1927 horizontal datum has been proven obsolete by subsequent, more accurate measurements. After many painstaking calculations, for 235,000 horizontal control stations and 500,000 vertical control stations, it is to be replaced by the new North American Datum (NAD-83), completed in 1985.[25] The National Geodetic Survey is also readjusting the nation's vertical control network, currently based on the Sea Level Datum of 1929. The addition of over 625,000 km [390,000 mi] of level lines since the 1920s is a major impetus for this readjustment, to be completed in 1988.[26]

Aerotriangulation and Stereoplotting

In the early twentieth century, photography and the airplane changed markedly the process of making a base map. The "bird's eye view" provided by a generalized picture of an area has always been a prime asset of maps, which, in a sense, have allowed the user to see over fences, around buildings, and beyond hills. Aerial photographic surveys have an advantage over terrestrial surveys in that taking a picture from an airplane and transferring ground detail onto a map is much simpler than the precise surveying and sketching of terrain features required in plane table mapping. Besides, a photograph provides far more terrain data than a map, an advantage readily recognized by military strategists. In less hectic circumstances, though, photography is used more as an aid to mapmaking than as a substitute for maps.

Aerotriangulation illustrates one way in which aerial photographs can expedite surveying and mapping. Consider three truly vertical aerial photographs, taken with the optic axis of the camera truly perpendicular to the horizontal so that the film was truly horizontal at the time of exposure. The center of a photograph, or *principal point,* thus represents the *ground nadir,* the point on the terrain directly below the camera at the time the shutter was snapped. In this example the three photographs overlap so that each photograph's principal point appears on the other two photos (Figure 3.10, left). Together these three principal points form a triangle, with one measurable angle on each photograph between the two

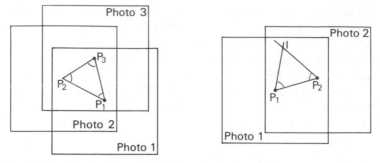

FIG. 3.10.   Overlapping aerial photographs can be used to measure angles for triangulation (left) and to fix additional control points by intersection (right).

radial lines from its center to the *transposed principal points*. Measurements of these angles on the photographs can be substituted for measurements taken in the field with a theodolite.

Phototriangulation can also be used to tie together points not at a ground nadir. If the distance is known between the ground nadirs of two overlapping photographs, the relative position of a third point, in the area of overlap, can be found by *intersection*. The line between nadir points is the known base of a triangle, two angles of which can be measured as shown in the right half of Figure 3.10. Moreover, if three control points with known coordinates appear on a single photo, the relative location of this photograph's ground nadir can be determined by *resection*. First, radial lines outward from the principal point to each control point are carefully drawn on a transparent overlay (Figure 3.11, left). Placed on a map showing the true relative locations of these control points, this overlay is then shifted by trial and error until each radial azimuth passes through the marked position of its corresponding control point (Figure 3.11, right). The point of convergence of the three radial lines fixes the location of the photo's ground nadir

A single vertical aerial photograph is a perspective view, not a planimetric map—it does not show true relative horizontal position. Angles may be measured accurately only at the principal point and unless the terrain is flat, relative distances between points are usually distorted. These limitations result from the conver-

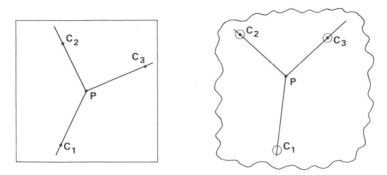

FIG. 3.11. Radial directions from principal point to three control points are plotted on an overlay (left) to be used in locating the ground nadir by resection on a map (right) showing the correct locations of the control points.

gence toward the camera's lens of lines of sight from the ground below. A true planimetric map would have points A and B in Figure 3.12 appear as if they were located at A' and B', on a horizontal *datum plane*. That is, their images on the photography would occur at a' and b', rather than at a and b. Note that point A, lying above the datum, has its image a displaced radially outward from its planimetric position a', whereas point B, below the datum, has an image b displaced radially inward from its planimetric position b'. This phenomenom, called *radial displacement,* precludes the accurate measurement of distance between most points on an aerial photograph. Relief displacement, which varies with elevation above or below a reference plane, is more extreme for points near the edge of the photo. If the format size is held constant, relief displacement can be reduced by either of two strategies for narrowing the angular field of view: increasing the flight height **h** or using a lens with a longer focal length **f**.

Relief displacement, which restricts the usefulness of the single photo, becomes an important asset when two photos overlap. The airplane taking separate photos over a pair of not-too-distant ground nadirs confers the depth percention of binocular vision. With but one eye we can judge depth of field only by experience with the relative sizes of objects, whereas with two eyes we can estimate the size of an unfamiliar form. Depth perception depends on the

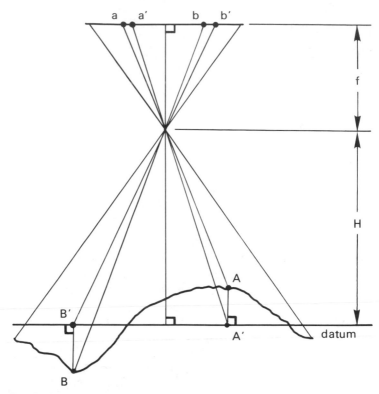

FIG. 3.12.   Geometry of a single vertical aerial photograph showing radial displacement, focal length **f**, and flight height **h**. Image of point A above the datum
plane is displaced outward from a' to a, whereas image of point B below the
datum is displaced inward from b' to b.

ability of the eye-brain system to detect subtle differences in "parallax," the change in position when an object is viewed from a
different direction. Suppose, for instance, we view a wooden dowel
sticking directly upward from the floor. If each eye operated like
a little camera, simultaneous pictures taken of the floor would
show images for the top of the stick a slight bit farther apart than
the pair of images for the bottom of the stick. That is, on our
cerebral snapshot the distance between the two images of the top

of the dowel will be greater than the corresponding distance between the two images for the bottom of the dowel. Figure 3.13 shows an equivalent situation for a flag pole viewed from two aerial cameras. Note that the parallax distance for the top of the flag pole is greater than the parallax distance for its base. We can view these photos with a *stereoscope,* one eye examining each photo, and our *stereoscopic vision* would enable us to convert the small *parallax difference* into an elevation difference so that we would recognize the pole as tall and vertical. On the individual photos it merely resembles a long, thin pole lying on the ground.

This parallax difference can be measured. Look at the upper half of Figure 3.14. For each section focus one eye on the left diagram and the other eye on the right diagram. Relax. Now try to fuse the two images. If you are a person with normal vision, you should be able to see a three-dimensional image without a stereoscope. The figure will resemble a deep pit with four steep sides. Repeat the process for the lower half, which will be seen as a pyramid. Note that in this example the parallax difference between the center and any corner is noticeable and measurable. A greater parallax difference would suggest a deeper pit or a taller pyramid.

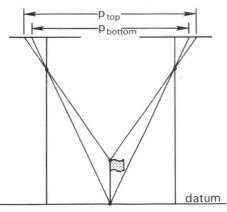

FIG. 3.13.   Geometry of a pair of overlapping aerial photographs showing the parallax difference between the top and bottom of a flag pole.

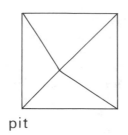

pit

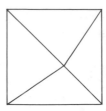

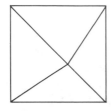

peak

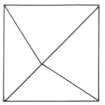

FIG. 3.14.   Simple geometric figures for pit (above) and peak (below) illustrate that, when positioned for stereoviewing, parallax is greater for lower elevations and smaller for higher elevations. Some readers might be able to see stereoscopic three-dimensional images by focusing one eye at the left part of each diagram and the other eye at the right part, and then attempting to fuse the images.

Simple stereoscopes that allow the user to view a three-dimensional model of terrain are useful in compiling maps because the compiler can readily identify most ground features. With the *stereoplotter,* a more complex instrument, a photogrammetric technician can trace the true planimetric paths of rivers, roads, power lines, and other linear features. He can also trace *contours,* lines through points at the same elevation that represent the shape of the land surface (Figure 3.15). Since the mid-1930s most topographic maps have been produced photogrammetrically. Other, less efficient methods for measuring elevation differences include trigonometric leveling with a transit and stadia rod and the use of an aneroid barometer to measure an elevation difference by the difference in atmospheric pressure.[27]

Even stereoplotting takes time. To prepare a manuscript map of contours, roads, and other features for an average 7.5-minute,

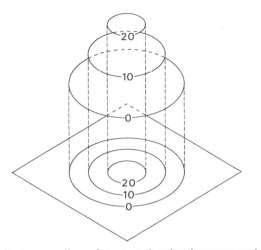

FIG. 3.15.   Contours are lines of constant elevation that represent the form of a land surface.

1:24,000- or 1:25,000-scale quadrangle map can require well over 200 person-hours for stereocompilation alone.[28] For areas with low to average relief, stereocompilation can be much more efficient with the *orthophotoscope,* developed in 1959 by Russell Bean of the U.S. Geological Survey.[29] The Gestalt Photomapper, an automated version of Bean's manually operated orthophotoscope, scans each photo in a stereo pair. The photos are treated as if decomposed into tiny squares organized according to a grid of rows and columns (Figure 3.16). The orthophotoscope senses the gray levels of individual grid cells and attempts to match a group of cells on one photo with a group of cells representing the same feature on the other photo. When a match is found, parallax distance is measured between groups. With a formula based upon the parallax measured for places with known elevations, the parallax distance between the correlated groups of cells is converted into an elevation estimate and a horizontal position for the center of the feature. The horizontal position estimate is adjusted for relief displacement. These (X, Y, Z) coordinates may be accumulated as data for a computer program that produces (1) a contour map, or (2) a *digital elevation model,* with terrain elevations

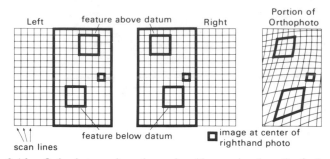

FIG. 3.16. Orthophotograph can be produced by copying the cells of a fine grid on an aerial photograph so that each grid cell is repositioned to compensate for radial displacement due to relief. Orthophotograph is produced for the overlap area of a stereopair by (1) correlating images for the left- and right-hand photos, (2) computing the parallax difference in order to estimate elevation relative to the datum, and (3) copying the image in its planimetrically correct position, with radial displacement removed. Image at far right shows center point and sample features above and below the datum plane repositioned from their displaced locations on the right-hand photo.

for the centers of cells on a somewhat coarse grid.[30] From a negative of one of the photos, the orthophotoscope can also produce a new photo on which each tiny square of the original image is printed in its planimetrically correct position. The resulting *orthophoto* has no relief displacement and can be highly useful in compiling a new or revised base map. With some feature labels and a coordinate grid added for user convenience, orthophotos are also used in making *orthophotoquadrangle maps,* quick temporary substitutes for a planimetric base map as well as useful planning tools in their own right.[31]

The geometry of aerial photographs is not as simple as the diagrams presented here might imply. "Vertical" photographs usually are not truly vertical—most contracts for aerial photographs, for instance, allow *tilt,* the angular deviation between the optic axis and a true vertical line, to be as great as three degrees. Yet, with proper ground control, small amounts of tilt can be removed photographically by a process called *rectification.* Because of small amounts of atmospheric turbulence, the aircraft cannot fly a perfectly level, perfectly straight flight line, and there are other distortions to the stereoscopic model as well. The photogrammetrist must contend with the *relative orientation* of the two photographs to each other as well as their *absolute orientation* to control points on the ground.

The adjustments needed to compensate for these distortions can be made on a stereoplotter by trial and error or mathematically with a digital computer. In *analog photogrammetry,* also called *instrumental photogrammetry,* the photogrammetrist removes distortions to the stereoscopic model by fine-tuning the various projectors on the stereoplotter. When the model is fully adjusted, he can view a three-dimensional scale model of the terrain and measure elevation differences to the nearest meter [3.3 ft] or less.[32] When the objective is to plot contours with an orthophotoscope or to estimate from existing control points the horizontal and vertical positions of other prominent features, *analytical photogrammetry* might provide a more efficient solution.[33] The photogrammetrist records the positions of corresponding images of the relevant points with a *comparator* and enters these data into a computer. A least-squares procedure estimates horizontal and vertical position. Analytical photogrammetry can also compute the adjustments needed to remove distortion on a stereoplotter. Through computer service bureaus or inexpensive minicomputers, even small photogrammetric mapping firms can take advantage of the efficiency of what once was merely an interesting but impracticable academic theory.

Inertial Positioning: Motion Indicates Distance

Aerotriangulation cannot fully eliminate the control surveyor's field work. Control monuments are small metal discs mounted in the ground, on the sides of buildings, or in other secluded locations to minimize disturbance by accident or vandalism. Not readily visible from the air, their locations must be marked for use in photogrammetry by large X-shaped targets, a meter or more across. If not centered on a control marker, these targets must be accurately positioned by field surveys and given new, carefully computed coordinates. Additional ground markers are often needed at locations well removed from the geodetic network. These and other field operations might be performed more rapidly and economically with an *inertial positioning system,* a comparatively new instrument that may eventually be as significant as electronic distance measurement and satellite positioning.

Based upon the gyroscope-controlled navigation systems of modern missiles and jet aircraft, an inertial positioning unit can be transported by helicopter or jeep from a known control point to a number of new control points. It records as it moves the accumulated displacements to the east (X), to the north (Y), and vertically (Z). The coordinates of a new control point can be estimated by simply adding the accumulated directional displacements to the coordinates of the control point at the origin of the traverse. For example, a new X coordinate, or easting, might be computed by adding the accumulated displacement 5,483.1 m from the easting for an origin, farther west, of 427,981.5 m to yield 433,464.6 m.

The displacements are accumulated separately for each of the three mutually perpendicular directions by adding the small displacements, positive and negative, experienced over many, very short time periods. Each small displacement is computed by multiplying the velocity in the specific direction by the time interval, about 17 milliseconds [.017 sec]. Velocity is computed from the accumulated difference in acceleration, which is measured directly, for each direction, by an accelerometer. The concept is similar to that of a pendulum: if a weight is mounted on the end of a thin flexible rod, the rod will bend to resist motion as the unit accelerates (Figure 3.17, left). The amount of acceleration is proportional to the displacement of the weight.

Accelerometers measuring acceleration in the X, Y, and Z directions need a stable platform. The required stability is provided by two gyroscopes mounted at right angles to each other. A gyroscope is a circular *rotor* spinning rapidly about its spin axle (Figure 3.17, right). It resists any movement that changes the alignment of the spin axle; in the 1920s, ocean vessels used large gyroscopes as stabilizers to reduce rolling. A spinning top is a crude gyroscope—attempt to push its axle to an inclined position, and it will right itself.

Rotor and spin axle are mounted within a *gimbal,* a surrounding ring containing the bearings for the axle. The axle can thus rotate freely in the plane of this ring. Addition of a second gimbal con-

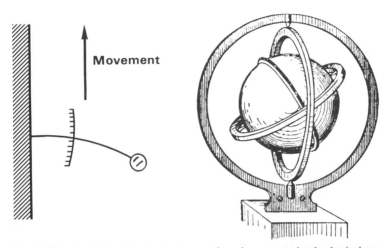

Fig. 3.17. Inertial positioning systems are based upon two simple physical systems. An accelerometer (left) is a weight at the end of a flexible rod with a sensor to measure acceleration by the displacement of the rod when the platform is moved. This acceleration is recorded for a continuous series of short time intervals and is used to estimate the total distance moved. The gyroscope (right), with a rotor spinning on an axle within two gimbals, resists shifts in orientation and is used to stabilize the platform so that separate accelerometers can measure displacement in the X, Y, and Z directions.

necting the first to the base allows the gyroscope to spin freely, with its axis always pointing in the same direction. In an inertial positioning unit, a servomechanism using a gyroscope to detect movement corrects the orientation of the platform containing the accelerometers.[34]

Unlike EDM instruments, inertial positioning systems currently are beyond the reach of the average land surveyor. At a 1981 price of $800,000 for the basic unit—the specially equipped jeep or helicopter is extra—few control surveyors are contemplating such a purchase. Yet the accuracy of these systems might justify for some projects their rental at rates of $100,000 per month. Although accurate to no more than about 10 m [33 ft] when first used in 1966, inertial positioning instruments are now accurate to within a decimeter and are appropriate for second- and third-order surveys.[35] They can be cost effective as well: the National Geodetic Survey, for example, saved two-thirds of the cost of conventional methods by using inertial positioning to establish 92 second-order control stations along a 480-km [300 mi] section of the Gulf Coast.[36] Control surveying is approaching that ultimate cartographic triumph of never, for any reason, having to leave the office.

Geodesy as a Public Utility

A nation's geodetic control network is not unlike its highway network. Both facilities are geographically extensive; both require careful coordination and continual maintenance. Because direct user charges are troublesome as well as costly to assess, both networks are public-sector obligations, supported largely from general revenues. In contrast, other public utilities, for example, public transit and metropolitan natural gas delivery systems, provide a service that is more easily metered, and which thus can be wholly or partly supported on a direct fee-for-use basis and be privately owned as well. Despite the future possibility of a direct charge for each use of geographic information and streets and roads, the geodetic control and road networks will continue to be government owned and maintained.[37] Both serve military as well as civilian goals, and both are basic to regional and national economic health.

To pursue this comparison further, both control surveying and road building have advanced greatly in the past century. Both networks have grown denser as well as more extensive. Because of improved engineering they have also become better—geodetic control more precise and highways faster if not safer. This improved technology has required still greater coordination and public investment, and has increased government involvement still further. Both the control net and the road net are fundamental elements of the public works infrastructure.

In the United States and other capitalist countries, influential lobbies of users, contractors, and equipment manufacturers agitate for growth, increased business, and continued improvement. To be sure, the "surveying and mapping lobby" pales in comparison to the size and clout of the highway lobby, and the immediate users of geodetic control are more likely to be federal and local government agencies than private surveyors. Yet both mapping and road building are influenced at least as much by political considerations as by progress in science and engineering. Technology advances in the channels opened by policy makers.

CHAPTER 4

Aerial Reconnaissance
and Land Cover Inventories

Map users must be told the what of a landscape as well as the where. Property surveys and the preparation of base maps, discussed in the previous chapter, involve measurement of distances, angles, and other geometric properties, and their producers customarily have backgrounds in engineering. Other measurements are needed to provide the details of land cover and terrain—to flesh out the skeleton, so to speak. This filling-in of detail involves both interpretation and measurement. The orbiting satellite and the computer have expanded the role of numerical measurements in geographic analysis, and geography is catching up with surveying and geodesy as a user of electronic technology. Whereas the previous chapter described how the Electronic Transition is making maps more positionally accurate, this chapter shows how electronic technology is making them richer in content as well.

To be fully use*ful,* terrain data must be made use*able.* It is not sufficient merely to collect information about a landscape and to provide interpretations; the data, in raw or refined form, must be organized for further measurement and interpretation as well as for convenient retrieval. This chapter addresses not only the collection and initial interpretation of terrain data, but also the role of digital cartographic data banks and computer procedures for measurement and analysis. Just as the computer can replace the paper map as a means of wayfinding, as discussed in Chapter 2, the digital data base can replace the library of paper maps as an archive of geographic information.

Topographic Mapping: A Base for Further Detail

Economic development requires some knowledge of a territory's geography. In early nineteenth century America, the expeditions of Lewis and Clark and other teams of explorers provided much of this knowledge, albeit in an unsystematic fashion. Somewhat later the land companies and railways penetrating the interior produced maps in the process of selling farmsteads, planning routes, and designing bridges. Plat maps, based on the Public Land Survey System, were prepared by county clerks, who recorded land ownership and assessments (Figure 4.1).[1] Private publishers often used these public documents for compiling county atlases showing the names of land owners as well as property lines (Figure 4.2). Distances were measured up and down hills and over winding, rough roads with a surveyor's *perambulator,* a large wheel of known circumference pushed ahead like a wheelbarrow (Figure 4.3). Horizontal accuracy was less a concern than salesmanship, and the "surveyor" traversed the countryside to impress and take orders as well as to measure and take field notes.[2] Personal vanity was a driving force behind the county atlas: the more prosperous land owners could usually be persuaded to purchase a copy or two of an atlas that contained an engraved picture of their farm or estate. The additional charge for sketching and engraving guaranteed a flattering picture of house and grounds, enhanced the self-esteem of the farmer, and enriched the publisher and his operatives.[3]

As the nation grew, the need arose for more detailed maps, covering larger areas and conforming to a widely accepted standard of accuracy. Henry Gannett, Chief Topographer of the U.S. Geological Survey, summarized the status of large-scale mapping in a presentation in 1892 to the National Geographic Society.[4] He contrasted the systematic national surveys of European nations and their maps of "uniform quality and character" with the United States' ". . . many partial surveys . . . made under independent authorities and of widely differing degrees of accuracy [with] the

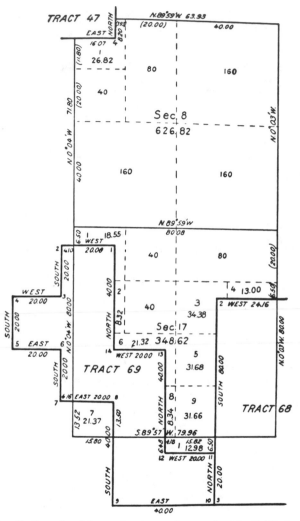

FIG. 4.1. Portion of a plat map showing section numbers for Public Land Survey townships. Tract boundaries, acreages, and some boundary lengths are also shown.

Source: U.S. Bureau of Land Management, *Manual of Instructions for the Survey of the Public Lands of the United States* (Washington: U.S. Government Printing Office, 1947), p. 437.

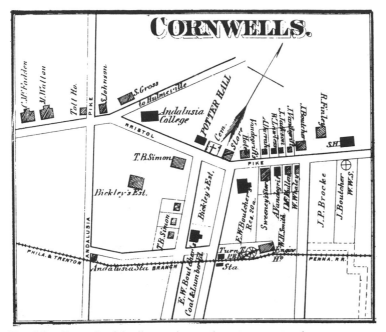

FIG. 4.2.   Example of detail on a nineteenth-century county atlas.
Source: Combination Atlas Map of Bucks County, Pennsylvania (Philadelphia: J. D. Scott, 1876), p. 53.

maps resulting therefrom differ[ing] in scale and value."[5] These surveys included the United States Lake Survey and other surveys of navigable waterways by the Army Corps of Engineers, surveys by the War Department and the Indian Bureau, the Pacific railway surveys, the Public Land Survey, and the surveys in the West of Clarence King, F. V. Hayden, and John Wesley Powell. The East, covered by scattered state and private surveys, was less well mapped than the West. Gannett noted:

> Large areas of the eastern and most densely settled portion of the country are dependent entirely for their maps upon [privately published] road diagrams of counties and upon railroad maps and profiles. Such is the condition of all in which no public land surveys have been carried on, excepting the areas surveyed by [federal agencies]. Thus, New York has no other maps besides these road

FIG. 4.3.   A surveyor's perambulator, or wheelbarrow odometer, for measuring distance.

Source: A Manual of the Principal Instruments Used in American Engineering and Surveying, Manufactured by W. and L. E. Gurley, Troy, N.Y., U.S.A., 29th ed. (Troy, N.Y.: W. and L. E. Gurley, 1891), p. 225.

diagrams, excepting 2,000 square miles mapped by the United States Geological Survey and some trifling additions by the United States Coast and Geodetic Survey, while Pennsylvania is almost as poor in information regarding its topography.[6]

The U.S. Geological Survey, founded in 1879, initiated a systematic topographic survey of the nation in 1882. By 1892, 703 map sheets had been surveyed and 600 of these engraved. These sheets covered 291 quadrangles extending over 15 minutes of latitude and 15 minutes of longitude at a scale of 1:62,500 and 352 30-minute quadrangles at a scale of 1:125,000. Sixty sheets at 1:250,000 were also produced. During its first 10 years, the Geological Survey's Topographic Division had surveyed 1,700,000 km² [650,000 mi²], over 20 percent of the nation's land area.[7]

Like most centrally administered programs in a federal republic, topographic mapping must respond to regional as well as national needs. Since 1885, the Geological Survey has responded to local demands through a program of federal-state "cooperation" in which the Survey performed the work and the state paid half the cost of survey, compilation, and drafting.[8] Although other federal agencies, such as the military and the Tennessee Valley Authority, have supported base mapping in various parts of the country, from time to time and in accord with their own needs, most of the nation's topographic archives can be traced to fifty-fifty cooperative funding. Figure 4.4 demonstrates the effectiveness of one state's influence as a "cooperator" in completing topographic coverage within its borders: for Pennsylvania between 1900 and 1932 the temporal pattern of area surveyed is very similar to the state's cooperation, measured in dollars not adjusted for inflation.[9] Pennsylvania perceived an economic advantage to topographic mapping—as did many other states with significant mineral resources—and overcame a slow start in the late nineteenth century to complete its own coverage before 1950, when barely half of the country had been surveyed for large-scale maps (Figure 4.5).

Technological innovation, especially the development of photogrammetric techniques, which could extract topographic information from aerial photographs, also had a significant influence

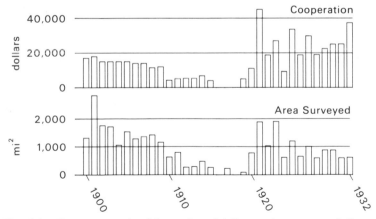

FIG. 4.4. State "cooperation," in unadjusted dollars, and area surveyed: Pennsylvania, 1900–1932.

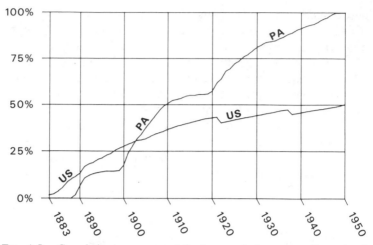

FIG. 4.5. Cumulative area surveyed for large-scale base maps: Pennsylvania and the United States, 1883–1950. Two reductions in cumulative area for the United States reflect adoption of a more stringent standard of accuracy.

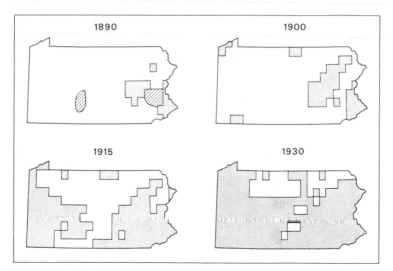

FIG. 4.6. Expansion of topographic coverage of Pennsylvania, 1890–1930. Diagonally shaded area on map for 1890 was not mapped by the Geological Survey.

on the evolution of Pennsylvania's topographic coverage. Topographic surveying proceeded slowly yet steadily: between 1890 and 1930 base map coverage expanded from a few scattered regions to over 80 percent of the state (Figure 4.6). By 1946, the topographic base included all but five 15-minute quadrangles, in the more remote, northern part of the state (Figure 4.7). Maps based on "modern" photogrammetric techniques, rather than older, less accurate plane table surveys, were more common in the north,

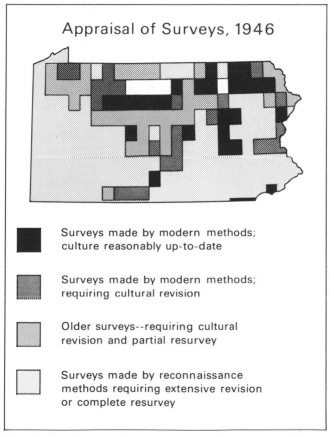

FIG. 4.7. Status of topographic map coverage for Pennsylvania, 1946.

which was mapped later than regions with more people, more coal, and more farms. Much of the remainder of Pennsylvania's coverage needed extensive revision or a completely new survey.

The Second World War had emphasized the need for accurate, detailed maps. It had also fostered both the training of topographic engineers and photogrammetrists and the continued improvement of instruments and procedures. Civilian mapping efforts inherited much of this infrastructure and rapidly shifted after the war from a 1:62,500-scale, 15-minute quadrangle to a more detailed, 1:24,000-scale, 7.5-minute quadrangle sheet (Figure 4.8). Photogrammetric methods enabled this new map series to expand rapidly, to over half the state by 1965 and to all of Pennsylvania by 1980. Many 7.5-minute quadrangles have been revised one or more times since their original resurvey, especially in areas of rapid landscape change, and very little of the state's base map is more than 10 years out of date (Figure 4.9). Pennsylvania's topographic data base, like those of most developed areas, now resembles a mature organism, with the continued, selective replacement of aging cells.

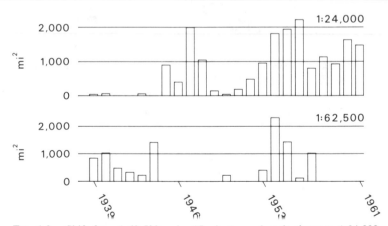

FIG. 4.8. Shift from 1:62,500-scale, 15-minute quadrangle sheets to 1:24,000-scale, 7.5-minute quadrangle sheets, 1939–1961. Histogram bars represent square miles covered by completely new or revised maps.

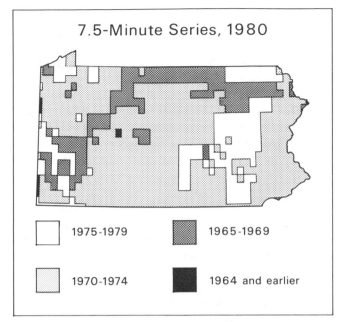

FIG. 4.9.   Pattern of publication dates for 7.5-minute topographic series, 1980.

Aerial Surveying and Photo Interpretation

Aerial surveying includes both aerial photogrammetry and air photo interpretation. If photogrammetry can be said to have simplified the "where" of mapping, air photo interpretation has most certainly simplified the "what." Features that otherwise would have to be observed on a slow, expensive sweep across the terrain by a surveying party can now be spotted rapidly in a climate-controlled office by a trained interpreter, who readily transfers their positions to the planimetrically correct base map developed by her technological kinsman, the photogrammetrist. Although some features might require a field check, the interpreter uses her knowledge of the phenomenon in question—geology, crops, soils, land use, military maneuvers, and the like—to decode the photo's contrasting tones, textures, sizes, and shapes. One of her principal tasks is "reading" the photo's grainy graytones and providing, in turn, a new map with specific features enhanced by standardized, stylized cartographic symbols.

The map prepared by the photo interpreter is a good illustration of the difference between the thematic map of the scientist and the general, or base, map of the topographer.[10] The photo interpreter is a specialist, who may be aided further by a *key* prepared to guide the classification of features related to a specific theme. Her training in mapping techniques commonly is minimal in comparison to her training in the subject matter mapped. Her concentration focuses upon specific spatial patterns and associations from which she is trained to derive meaning. Her map, if one is prepared at all, commonly requires specialized knowledge for its own interpretation, or at least a fairly detailed explanation of its symbols. The photogrammetrist's map is merely the cartographic skeleton to which the interpreter contributes the geographic or scientific flesh.

As its name implies, aerial photography has two components: an aerial platform that affords an overview of the terrain and an image recording device, or sensor. These components were invented and developed independently, and both have advanced considerably since the days of the Wright Brothers and George Eastman. Today, in fact, more territory is mapped more frequently by orbiting satellites and electronic, nonphotographic image recorders than by fixed-wing aircraft carrying cameras with glass optics and photographic film. And various combinations are possible as well: manned spacecraft have used photographic systems and simple aircraft carry sophisticated electronic terrain sensors.

The earliest aerial platforms were not airplanes but balloons. In 1858, Gaspard Felix Tournachon, a French photographer with ambitions of producing a topographic map, photographed a village near Paris from an altitude of several hundred meters. Nineteen years earlier, in 1840, the French had hinted at the prospect of aerial photography, albeit as a spoof of photography in general. A cartoon depicting the taking of ground photos from a balloon bore the caption, "Daguerreotypomania." Early photographs were called Daguerreotypes, after Louis Daguerre (1787–1851), acknowledged as the co-inventor of photography together with an-

other Frenchman, Joseph Niepce (1765–1832). Photography improved steadily through the mid-nineteenth century with the
development of more sensitive emulsions, sharper lens systems
with less distortion, and lighter cameras.[11]

A variety of platforms were tried. Balloons were either free,
like the hot-air balloons in use today for recreational flights, or
captive, bound to the ground by ropes for guidance and ready
retrieval. Captive balloons were preferred for military reconnaissance—the intelligence gathered was of little use if the mission
landed behind enemy lines. The armies of various nations exper-
imented with unmanned kites, and even carrier pigeons with small
timer-operated cameras strapped to their chests.[12] Kites, of course,
required a moderate wind in the direction of the territory to be
photographed, and homing pigeons had to be released on the far
side of the target area from their home coop. Positioning and
recovery was also a drawback to experiments with the primitive
rockets of the early twentieth century. Although a compressed-air
rocket could lift a camera almost a kilometer into the air, with a
parachute providing a slower return to the ground, the resulting
few low-altitude, small-scale photographs, blurred and poorly
centered, hardly justified further serious development.

The first photographs from a fixed-wing aircraft were for entertainment, not mapping, and were taken almost six years after
Orville Wright's historic 12-second, 40-meter [120-ft] flight on
December 17, 1903, over the beach at Kitty Hawk, North Caro-
lina. On April 24, 1909, his brother Wilbur flew an improved
plane, patented in 1906, over Centocelli, Italy, while his passen-
ger filmed the ground with a motion picture camera. Despite this
surprisingly slow start, aerial photography grew steadily over the
next decade, with the further development of aircraft and the need
for military reconnaissance during World War I providing the
greatest impetus. Promoted largely as a weapon, the airplane could
also scout the positions of enemy troops, fortifications, and supply
lines, and return quickly with valuable film, which could be processed rapidly and then studied intensively by military planners.
Such attempts at evasion as camouflage and the use of decoys
attested to the tactical value of this early aerial imagery.[13]

Except perhaps in Germany, photo interpretation developed

slowly after World War I. American interest in military matters waned considerably after the Armistice, but small survey firms recognized the potential of the aerial camera for civilian mapping. One entrepreneur, Sherman Mills Fairchild (1896–1971), not only founded a company that in its first year, 1922, did $600,000 worth of mapping for such cities as Boston, Kansas City, Newark, and New York, but also made major improvements in both aerial cameras and closed-cabin airplanes.[14] Municipalities interested in expanded public works, more complete tax assessment, controlled growth through zoning, and industrial development needed both base maps and land use maps. Articles in geographic, geologic, ecological, planning, and other professional journals made scientists, engineers, and public administrators aware of the wide range of uses of photo interpretation. In 1934, the American Society of Photogrammetry formed to advance both photogrammetric mapping and photo interpretation. The choice of Washington, D.C., as its headquarters reflected an increased use of aerial imagery by government agencies for mapping, planning public works projects, and monitoring agricultural and timber production.

Yet the boost to photo interpretation by Roosevelt's "alphabet agencies" of the 1930s was far less significant an impetus than World War II. General Werner von Fritsch, chief of the German General Staff, stated in 1938 that "the nation with the best photo-reconnaissance will win the next war."[15] The United States, far behind Germany in the techniques of aerial intelligence—but fortunately not behind Japan—initiated a crash program to train military interpreters, to expand the production of cameras and aircraft, and to improve its imaging systems. The Air Force established a research laboratory at Wright Field, Ohio, with Colonel George W. Goddard as its director. Working with the Eastman Kodak Company, Goddard's laboratory tested and developed films that could detect camouflage.[16] Other significant technical advances arising from the war effort include nearly distortion free lenses; improved camera mounts that absorb aircraft vibrations; film drives to compensate for changing aircraft velocity; rapid, reliable shutters; improved portable stereoscopes and stereoplotters; and various navigation aids to assist the pilot in following the planned flight lines.[17] Even more significant, though, has been the training

and practical experience given many thousands of photo interpreters. This expertise provided the momentum for numerous land
use, geologic, engineering, agricultural, and other applications of
photo interpretation in still another demonstration of the transfer
of mapping technology from military to civilian uses.

Remote Sensing and Orbiting Platforms

The Cold War of the 1950s and the post-Sputnik arms race of the
1960s and 1970s added a new urgency to the search for more
effective methods of aerial surveillance. Image sensing and recording techniques had developed far beyond the reconnaissance
cameras of the 1940s, and by the mid-1970s pictures of the Earth
were routinely transmitted from orbiting satellites to a worldwide
network of receiving stations. In colleges and universities, courses
on "Remote Sensing" rapidly replaced or augmented courses in
"Air Photo Interpretation" to reflect not only an increased range
of platform altitudes but also imaging systems that no longer were
solely photographic.

Camouflage detection film, a product of the war effort of the
1940s, was one of the more significant developments leading to
the emergence of remote sensing from photo interpretation. Military tacticians needed to distinguish on aerial photos real vegetation from potential targets painted to mimic the color, tones, and
textures of plant cover. The key to detecting camouflage is that
healthy green vegetation "looks" green because the green light is
reflected by the plant more than the blue and red light. Therefore,
if we observe only the visible part of the electromagnetic spectrum, with either our eyes or standard color film, color is useless
as a cue for differentiating healthy vegetation from the green camouflage paint, which reflects almost exactly the same amounts of
green, blue, and red light as natural foliage (Figure 4.10). Yet if
our eyes were sensitive to light in the near-infrared band—the
part of the spectrum with wavelengths just longer than those of
visible red—living plants would "look infrared" whereas green
paint would still appear green. Healthy vegetation reflects both
green light and infrared radiation, but the infrared is more prominent and dominates the green.

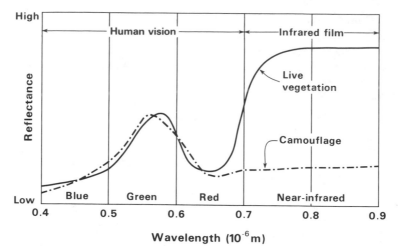

F<small>IG</small>. 4.10.   Spectral "signatures" of healthy green vegetation and green-painted camouflage are similar in the visible part of the spectrum, but clearly distinct in the near-infrared.

Color infrared (IR) film uses this concept and a "spectral shift" to provide a "false color" image that distinguishes camouflage from real vegetation by showing the former in blue and the latter in brilliant red. Standard color infrared film has three emulsion layers that are sensitive to green, red, and infrared light. During development the three layers receive yellow, magenta, and cyan dyes to cause a color or spectral shift in the photograph. This shift causes green objects to appear blue, red objects to appear green, and infrared objects to appear red.[18] The interpreter can now "see" the strong infrared reflectance of live vegetation, which contrasts sharply with the pale blue or weak brown of camouflage (Figure 4.11). In the civilian sector, color IR imagery finds its principal use in agricultural, ecological, and forestry studies.

Remote sensing measurement extends well beyond the visible and near-infrared bands. The first Landsat satellites, for instance, carried two sensors, both nonphotographic. The return beam vidicon (RBV) system used three cameras with television scanners to view portions of the Earth 185 km by 185 km [115 by 115 mi]. Each camera was sensitive to reflected radiation in a different part of the spectrum; designated channels 1, 2, and 3, these cameras

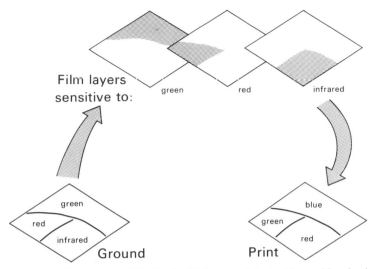

Fig. 4.11.   The spectral shift of color IR imagery detects objects with a dominant reflectance in the green, red, or near-infrared band but represents these objects symbolically as blue, green, or red, respectively.

individually were sensitive to reflected green, red, and infrared energy, respectively (Figure 4.12). Unlike standard television pictures, RBV images were recorded periodically rather than continually on a photosensitive surface. A television subsystem then scanned this surface, more slowly and precisely than in a conventional "live" broadcast, and transmitted the image to the ground as a video signal.[19] In contrast, Landsat's multispectral scanner (MSS) system sensed four bands, labeled channels 4, 5, 6, and 7, covering green and red with one band each and the near-infrared part of the spectrum with two bands (Figure 4.12). Like a common kitchen broom, the scanner would advance along a scan path of 185 km [115 mi] long, sweeping its scan from side to side perpendicular to the direction of travel, with each sweep simultaneously recording information for six scan lines 79 m [260 ft] wide (Figure 4.13). An oscillating mirror focused the energy reflected from the scan path onto sensors that measured the energy received for each band as an integer in the range 0 to 127, say, so that the image could be transmitted to Earth as a digital signal.[20] Measurements were made every 56 m [180 ft] along a scan line,

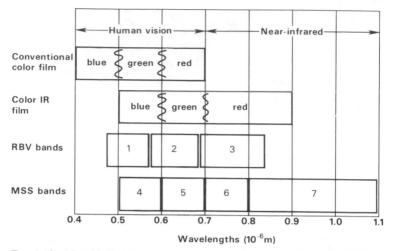

FIG. 4.12.   Sensitivity of the human eye, conventional color film, color IR film,
and Landsat RBV and MSS imagery.

so that the resulting image consisted of a grid of *pixels* 79 m by
56 m.[21] As made available on magnetic tape, these pixels are
organized into *scenes* covering about 185 km by 185 km with a
grid of 2,340 scan lines and approximately 3,240 pixels per line.

An innovative variation developed for the French SPOT (Satellite Probatoire pour l'Observation de la Terre) program is the
push-broom scanner, which, with no moving parts, is lighter and
inherently more reliable than Landsat's oscillating-mirror system.
The scanner's linear array detector is a line of as many as 10,000
microdetectors joined edge to edge. All detectors in the array can
be sampled almost simultaneously, in close succession, as the
scanner sweeps along its *ground swath* with the efficiency of a
push broom. This improved design is likely to be adopted by the
United States and other countries.

Engineers and earth scientists have designed many hypothetical
satellite sensing systems theoretically adapted to specific information needs. Among these are Mapsat and Stereosat, two spec-
ulative satellites designed to collect elevation information. Linear
array detectors in the sensors planned for Mapsat would be aimed
26 degrees fore and aft to provide images that could be viewed in

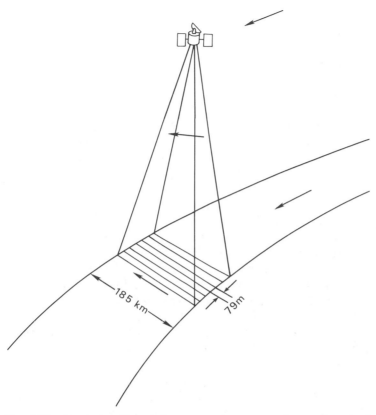

F<small>IG</small>. 4.13.   Landsat multispectral scanner system sweeps six scan lines simultaneously in a scan path 185 km [115 mi] long. Scan lines are 79 m [260 ft] wide.

stereo and used in an advanced orthophotoscope to extract elevation information. Mapsat is designed to provide elevation data for mapping at 1:50,000 with a 20 m [66 ft] contour interval. A similar concept is the basis for Stereosat, planned to yield stereoscopic imagery for geologic studies.[22]

Many applications require photographic prints. Reflected electromagnetic energy recorded by the sensors can serve as data for a printer that produces photographic images. Landsat's and other RBV and MSS systems have the spectral sensitivity of color IR film, and scenes from these sensors can be converted to color-composite photographic prints distinguishable from photographic color IR imagery only by occasionally obtrusive scan lines. Color and black-and-white prints are provided at scales of 1:1,000,000 and 1:250,000, and custom enlargements are possible as well. Each band has unique advantages for interpreting particular types of features, and many interpreters prefer black-and-white prints to those in color. Band 5, for instance, shows light-toned major highways very clearly, whereas Band 7 provides sharp land-water boundaries.[23] Because the ground is far below the sensor, there is very little relief displacement. Digital Landsat images are also available on magnetic tape for computer analysis and use with interactive image analysis systems.

Landsat-3, launched in 1978, six years after Landsat-1, extended the interpreter's vision still further along the electromagnetic spectrum by including a thermal, or far-infrared, band in its MSS system. This thermal sensor was to detect differences in surface heating, differentiate between healthy and stressed vegetation and detect differences in soil moisture. A resolution of 240 m [790 ft], instead of 79 m [260 ft], focused more thermal energy on the detector and promoted greater sensitivity to thermal radiation from the ground. Unfortunately, the thermal channel failed shortly after launch in 1978.

Thermal bands complement information obtainable from the visible and reflected infrared bands. For this reason, the Thematic Mapper, a sophisticated seven-band multispectral scanner launched on Landsat-4 in 1982, senses in visible, near-infrared, and thermal-infrared bands.[24]

Sensors may be active or passive. Like a standard camera without a flash attachment, Landsat MSS imagery is passive: it depends upon solar radiation reflected or re-radiated from the Earth. In contrast, active sensors generate their own energy, which is reflected by the target to be measured and recorded in turn by the sensor where it originated. Like a camera with an electronic flash, radar and other active sensing systems can be used at any time of day. Unlike photographic systems, these microwave systems can penetrate clouds, haze, rain, snow, fog, and smoke. Mapping can be scheduled at the convenience of the scientist and can respond

to the immediate requirements of the military planner, without the
need to wait for daylight or a cloud-free sky.

Radar is typical of most active sensing systems. Images are
formed by accumulating a two-dimensional record of radiation
backscattered from the land surface. Consider the plane flying to
the left of the building and clump of trees in Figure 4.14. At some
instant a radar antenna carried by the plane radiates high energy
pulses toward the building and trees. Because of its size and substance, the building reflects more energy back to the plane's receiving antenna than do the trees. Moreover, backscattered energy
generated in a particular pulse will be received back at the plane
sooner when reflected from the closer building than when reflected from the more distant trees. The lower part of Figure 4.14
illustrates the relationship between the strength and arrival time of
the reflected energy. Note that time, shown on the horizontal axis

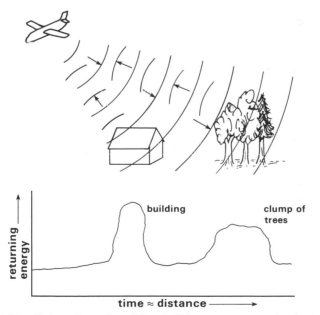

FIG. 4.14.   Radar pulses reflected from building are stronger and arrive back at
the sensor sooner than pulses returning from more distant, less reflective clump
of trees.

of the graph, is a surrogate for distance. Similar curves describing measured return energy and return time can be generated for parallel scan lines, usually on one but occasionally on both sides of the aircraft (Figure 4.15). These scan lines, perpendicular to the plane's flight path, form a raster image of scan lines, similar to Landsat MSS imagery. Analog-to-digital conversion and adjustment for the decline in scale with increased distance from the flight path yields a gridded image of pixels. A black-and-white photographic print can be produced with a film recorder from either an analog or digital representation of the radar image. This type of sensing system is called *side-looking airborne radar,* or SLAR.[25]

Radar imagery, which detects gross terrain features, is particularly useful for the rapid reconnaissance mapping of natural resources and geologic structures. It yields a different view of the terrain than do images based on heat or reflected light: textures on active and passive images of the same features often differ considerably. Radar brightness, the strength of backscattered energy received at the antenna, depends on a variety of characteristics: the slope of the land surface and its orientation to the radar antenna; the relationship between the wavelength of the incident microwave radiation and the roughness of the terrain; and the poorly understood physical properties of surface materials that determine inherent differences in radar reflectivity.[26] If the form of the land surface is represented by a digital elevation map, with elevation values recorded for pixels similar to those in the raster-mode radar

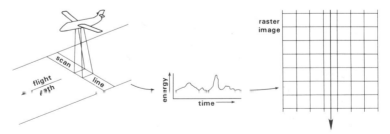

FIG. 4.15.   Scan lines sensed by side-looking airborne radar (SLAR) have a time-pulse strength curve that can be converted to a digital raster image with pixels similar to those of a Landsat scene.

image, adjustment may be made for topographic slope in order to detect differences in surface materials.[27] With its high altitude producing a nearly constant scale and a constant angle of illumination, an orbiting radar sensor carried by Seasat yielded radar imagery highly useful for mapping landforms in areas of low topographic relief as well.[28] In contrast, SLAR imagery taken from a fixed-wing aircraft in mountainous terrain often suffers from black, sharply defined shadows for land hidden from or facing away from the sensor. This is called *layover*.

The spacecraft's orbit and sensor characteristics determine the scale and areal coverage of the imagery. Sensors designed to map or inventory natural resources are often placed on satellites in circular, low-altitude orbits to provide a constant scale and fine spatial resolution. A *sun-synchronous* orbit, in which the orbital plane progresses in the direction opposite the Earth's rotation, can provide complete coverage of a large area, with periodic repeat coverage and consistent illumination of all scenes taken during the same season at the same latitude. The early Landsats' near-polar orbits swept the illuminated side of the Earth from north to south at about 9:42 A.M. local sun time. A scene was recorded every 18 days and successive ground swaths were about 25 degrees of longitude apart at the equator (Figure 4.16). Orbits may also be *synchronous polar*, with a polar orbit rotating with the Earth and following the same two opposing meridians in order to provide frequent coverage of the same place, or *synchronous equatorial*, with the satellite always far above the same point on the equator and making one complete revolution with the Earth every day. Synchronous polar and synchronous equatorial orbits are preferred for weather and intelligence satellites, so that continental atmospheric conditions or specific tactically important sites may be sampled frequently. High-altitude synchronous equatorial orbits are needed for communications satellites, which must be "visible" from a widespread group of ground antennas with fixed orientations.[29] The Space Shuttle is a particularly promising orbital platform, which can deploy and retrieve satellites as well as collect experimental data in its own low-altitude orbit.[30]

Spatial resolution, the smallest object on the ground that can be discriminated, is a fundamental characteristic of all remotely sensed

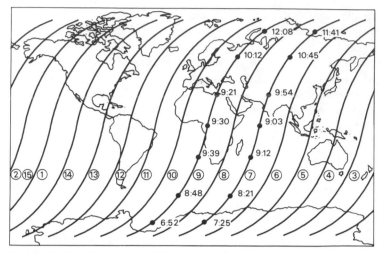

FIG. 4.16.   Landsat orbits the Earth 14 times a day in a near-polar orbit, with each orbit crossing the equator about 25 degrees west of the previous crossing. Typical local time variations are shown for two orbits. Orbit 15 is slightly west of Orbit 1 to provide coverage in the adjoining ground swath on the following day. This pattern is repeated every 18 days.

Source: Landsat Data Users Handbook, revised ed. (Reston, Va.: U.S. Geological Survey, 1979), p. 5-2.

imagery, but despite the simplicity of the concept, resolution is not easy to measure. The most common measure of resolution is the *instantaneous field of view* (IFOV), the side or diameter of a patch on Earth from which energy is received by a detector for a single reading. For Bands 4, 5, 6, and 7 of the MSS on Landsat-1, -2, and -3 the IFOV is 79 m [260 ft]. On Landsat-3 the RBV system IFOV is 40 m [130 ft] and the thermal scanner IFOV is 240 m [790 ft]. Landsat-4's Thematic Mapper employs an IFOV of 30 m [98 ft] for its visible and reflected infrared bands and 120 m [390 ft] for its thermal band. France's SPOT program uses a linear array detector with IFOVs of 20 m [66 ft] and 10 m [33 ft]. For intelligence satellites, of course, the IFOV is much smaller— sufficient, some say, to distinguish the dress of an individual on the ground as either civilian or military.[31] Blurring resulting from aberration in the optics and diffraction, as well as from the forward motion of the satellite and rotation of the mirror, may reduce the sensor's ability to detect small objects.[32]

Equally important is the contrast between a ground feature of interest and its surroundings. Even if narrower than the IFOV, roads and other linear features contrasting strongly with neighboring features can often be identified with ease: despite the averaging of energy reflected from pavement and roadside, the brightness recorded along a chain of pixels following a highway can be substantially different from that of surrounding cells. Yet a feature somewhat larger than the IFOV but not greatly dissimilar in reflectance from its environment may have its otherwise prominent reflectance diluted by its surroundings, particularly if divided nearly equally among four adjoining pixels (Figure 4.17). Intricate, irregular boundaries between contrasting land types produce a similar *edge-averaging effect,* and in urban areas the combined effect of two or more roofs of different pitch or material as well as roadways, sidewalks, driveways, tops of vehicles, grass, trees, shrubs, patios, and bare ground may yield a spurious averaged reflectance characteristic of none of the objects within the pixel— or, as it might be called in this case, the "mixel." Moreover, these

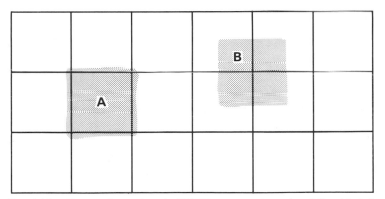

Fig. 4.17.   Features larger than the IFOV but contrasting only mildly with their surroundings may be distinguishable if they dominate a pixel (A), or unrecognizable if diluted by neighboring features when divided among several pixels (B). Cells sharing feature B will be only slightly different from their neighbors, and this difference may be indistinguishable from the overall variation among cells with this land cover.

averaged reflectances may well vary unsystematically with the relative orientation of scan lines and street grid, and produce a reflectance that is meaningful only if averaged further, say, for a 2 by 2 subgrid or a larger cluster of pixels. If a coarser resolution is appropriate for delimiting the extent of a certain type of land cover, such averaging can simulate the effect of a larger IFOV. Yet if the resolution needed is finer than that available, the analyst has little hope of obtaining greater spatial detail. The limitations and costs of data transmission and storage not withstanding, with satellite imagery the finest possible resolution is almost always preferable.

Image Analysis and Image Enhancement

Data recorded by a remote sensor and stored as numbers on magnetic tape become useful information only when linked to the ground, other digital data, and the analyst's experience with the region. The interpreter's first need is a classification system for land cover that accommodates both the goals of the project and the limitations of the imagery—its spatial resolution and spectral sensitivity. This classification system determines how detailed and informative a map might be produced from the digital imagery. Multiple areas on the ground represented by several contiguous pixels must be found for every category in the classification. These *ground truth,* or *training,* data are then used to assign other pixels with similar reflectance to appropriate categories. If the ground truth data are faulty, the interpretation will be flawed, and if adequate ground truth areas cannot be found, the classification will have to be revised.

Among the various strategies for classifying multispectral imagery, the *parallelepiped classifier* is the simplest.[33] The interpreter attempts to define a land cover type by bracketing the range of ground truth values for each spectral band. Figure 4.18 illustrates this principle for a simple, two-band classification based on Landsat imagery. A rectangle is defined around each land-cover type by the highest and lowest reflectances for each band. Three bands would require a three-dimensional scatter diagram and would yield for each category a true parallelepiped, with six sides, each

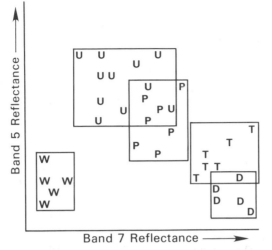

Fig. 4.18.   Rectangular regions defined by minimum and maximum ground truth reflectances represent land-cover types in a simple, two-channel parallelepiped classifier.

representing an upper or lower limit for one of the scanner's channels. Letters in the two-dimensional scatter diagram shown identify pixels as deciduous forest (D), turf or cut meadow (T), urban land (U), water (W), and freshly plowed field (P). Other pixels with pairs of reflectances within one of these sets of limits could then be assigned to the corresponding category.

This example illustrates two difficulties with the parallelepiped classifier. First, some land-cover types, such as the plowed field and urban categories, overlap. Moreover, training pixels might occupy only a small part of their parallelepiped's interior, especially if there is a pronounced linear trend, such as the positive correlation between Bands 5 and 7 for turf and cut meadow. In these cases stepped regions, in the form of a series of smaller contiguous or overlapping parallelepipeds, might reduce ambiguity and the probability of misclassification by defining more tightly and precisely the multivariate range of each category. Second, much of the area on the scatter diagram is not assigned to any land-cover type, with the possibility that many slightly unique, transitional areas would be assigned to a cumbersome "undefined" category. This uncertainty might easily be tolerated in highly focused research projects, for which the interpreter might need only to determine the areal extent of one or two well-defined land-cover types. For a general purpose map, though, extensive interactive experimentation with category labels and limits is often needed to reduce the number of undefined pixels to an acceptable level.

More advanced classification algorithms, based on probability theory, recognize trends and clusters in the data and can estimate the probability that every pixel in the scene belongs to each category in the classification. Pixels close to the core of a compact, distinct cluster can be assigned to that category with a very low probability of misclassification, whereas other pixels, not clearly belonging to a particular class, are allocated to the land-cover type to which they are judged most similar. Clustering algorithms can estimate the overall accuracy of the classification and even suggest that certain training pixels have been misclassified.[34] Accuracy estimates provided for individual categories may be useful in revising the classification to produce a more accurate, more meaningful map.

In general, an increase in the number of spectral bands may be used to extract additional systematic differences among the pixels and improve the accuracy of classification. [35] With remote sensing programs such as Landsat, which provide continual, periodic coverage over several years, additional bands can be added, in a sense, by merely integrating imagery for the same area but obtained during different seasons. Coniferous and deciduous forests, for instance, have pronounced seasonal differences, as do various field crops when sampled during different stages of development. Multitemporal imagery is highly useful in detecting changes in land use, in monitoring the spread of plant diseases, and in suggesting the need to revise more detailed maps.[36]

Clouds generally reflect far more energy than the land surface in both the visible and reflected infrared parts of the spectrum; in areas covered by clouds on one of the images, a very high reflectance is almost always the result of sunlight reflected from a cloud. Repeat coverage can also compensate for problems caused by scattered cloud cover. A relatively cloud-free scene might be obtained by assigning each pixel in a new image the lowest reflectance of the corresponding pixels in two scenes of the same area taken several days apart.

Proper registration can be a major obstacle in multitemporal analyses. Subsequent passes over the same ground swath by a satellite such as Landsat cannot guarantee a pair of perfectly registered scenes. Because of orbital perturbations and the idiosyncrasies of the sensor, the center of what is intended as the same scene may vary by as much as 37 km [23 mi] from the center of the imagery recorded 14 days earlier.[37] One solution is the removal, through *resampling,* of geometric distortions. Because of the Earth's rotation, each scan's pixels are offset slightly from those of the previous scan line, and the overall boundary of the scene is in fact a parallelogram, not a rectangle (Figure 4.19). Resampling can correct for this distortion too. Reflectance values from the original MSS image are used to estimate the correspond-

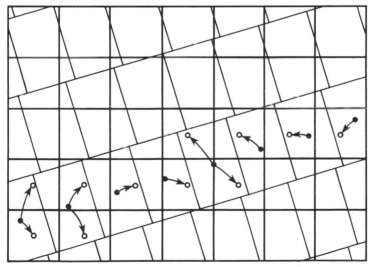

Fig. 4.19.   Resampling of original, offset grid of rectangular pixels to a new, geometrically corrected grid of square pixels may assign the reflectance of an original cell to more than one new cell.

ing reflectances for a new array with square grid cells aligned to conform to a standard coordinate system such as the Universal Transverse Mercator grid. Values for the new pixels are estimated by either (1) computing a weighted average of neighboring cells, so that smoothing diminishes the sharpness of the original imagery and effectively increases the IFOV, or (2) assigning each new pixel the reflectance of the nearest pixel on the original image, so that a recorded reflectance might well represent conditions as much as half a pixel away from its stated location (Figure 4.19). Classification algorithms for multitemporal imagery should consider reflectances in the neighborhoods of each pixel rather than treat each cell's reported value as independent of its neighbors'.[38]

Algorithms that compare neighboring pixels aid interpretation by either improving contrast or filtering out noise.[39] An *edge enhancement* procedure sharpens an image by heightening the contrast between a pixel and its neighbors. Locally low reflectances can be made lower and locally high reflectances higher by comparing a pixel's value with the average for its eight neighbors and doubling the difference (Figure 4.20, upper). Edge enhancement

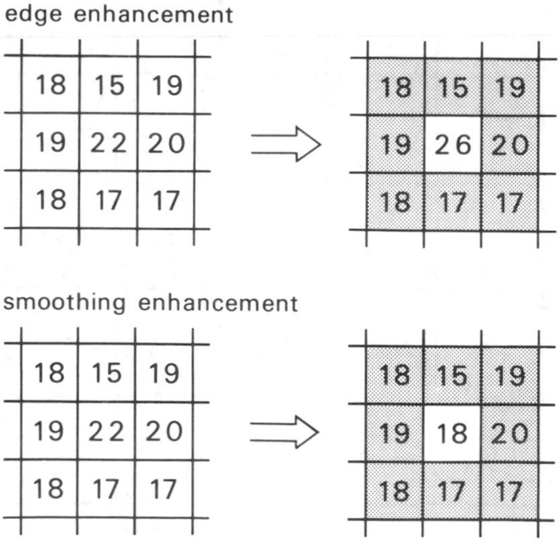

edge enhancement

smoothing enhancement

FIG. 4.20.   Local operators exaggerate pixel's difference from its neighbors in edge enhancement (above) and diminish this difference in smoothing enhancement (below).

compensates for the sharpness lost when a coarse IFOV dilutes differences between neighboring land covers with contrasting reflectances; pixels in the enhanced image receive brightness values closer to one of the original land covers than to their own average. Yet local sharpness is not always desirable, and a *smoothing enhancement* operator might be invoked to minimize distortions caused by locally high or low brightnesses unrelated to a broad regional pattern. These "spikes" are removed by substituting the average reflectance for the nine pixels within each 3 by 3 neighborhood for the raw image value at its center (Figure 4.20, lower).

Efforts to compensate for possible misregistration are important not only for multitemporal classification but also for overlay analyses linking remotely sensed data to existing geographic data bases. Known ground conditions such as elevation data may provide additional information useful in developing a more refined classification of land cover or in sharpening an image. Elevations recorded for the centers of grid cells can be used to estimate slope, which correlates with both vegetative cover and human activity— and which might function, in a sense, as an "additional band" for a more informative land-cover classification. Conversely, a grid of cells each assigned to a land-cover category and registered to a digital elevation model can be a valuable tool for regional planning and environmental analysis. Areas susceptible to severe soil erosion might be identified, for example, as cells with steep slopes and sparse vegetation. Similarly, acceptable sites for a new airport might be identified by overlaying topographic and land-cover data and searching for extensive tracts of uninhabited, level land. Natural resource data banks might also contain land use information interpreted from conventional, low-altitude aircraft, and soils information collected through an intensive and thorough field survey. Property ownership records and zoning restrictions might be included. Each addition of relevant data to a land information system increases the value of previously contributed, equally accurate data, which might then be related to a wider variety of geographic distributions and yield new and potentially useful insights.

Computers able to process rapidly large amounts of geographic data are particularly valuable to military planners, and even more so when airborne scanners, operating in real time, collect radar imagery of enemy territory. Periodic sampling and comparison of digital imagery can detect movement of enemy convoys. Data describing terrain and transport routes, as well as possible destinations and wooded areas that might afford shelter, can be useful in predicting an enemy's evasive behavior if attacked. Such systems serve strategic surveillance as well as tactical planning, and might be used to monitor agricultural production, industrial development, heavy construction, and other activities indicative of a potential enemy's problems and aspirations.

Far less ominous is the use of a computer to examine digital maps in order to make other, more refined displays for researchers, decision makers, and the general public. Most early computer-produced maps were *choropleth maps,* with the geographic patterns of percentages and ratios shown by area-shading symbols, light for low values and dark for high values. Choropleth maps are usually crude and often misleading because these percentages and ratios often vary considerably within the individual counties, states, or nations for which they are reported. When plotted for a state by county unit, for example, a choropleth map of population density makes the naive assumption that a county's residents are uniformly distributed within its borders. A geographic data system might read a digital land-use and land-cover map to identify uninhabited areas so that more reliable ratios of people to inhabitable land can describe the state's pattern of residential intensity (Figure 4.21).

Remotely sensed data and computer software can assist in the production of an even more refined graphic display of density, the dot-distribution map. Each dot represents a given number of persons, farm animals, or some other phenomenon, and the user perceives density variations from the relative spacing of dots. A computer can use a census data file to calculate the number of dots to be placed within each county and employ a digital land-cover map to determine the most plausible locations of these dots (Figure 4.22). Since 1969, the U.S. Bureau of the Census has used a computer algorithm and a land use data base to plot numerous national overview maps for data collected in its periodic agricultural censuses.[40] The increased availability of digital remotely sensed data can extend this approach to larger-scale maps

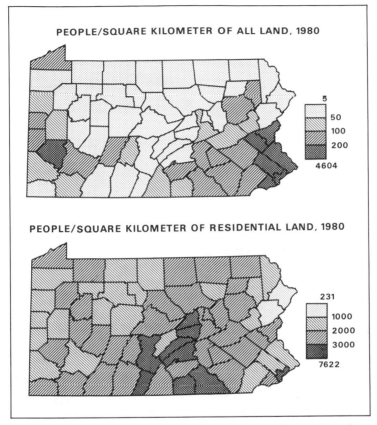

PEOPLE/SQUARE KILOMETER OF ALL LAND, 1980

5
50
100
200
4604

PEOPLE/SQUARE KILOMETER OF RESIDENTIAL LAND, 1980

231
1000
2000
3000
7622

Fig. 4.21.   Different choropleth maps result for Pennsylvania county population densities expressed as ratio of population to all land (above) and ratio of population to inhabitable urban land only (below).

of states and counties. Producing a dot map manually is a tedious, time-consuming task—and clear proof that human inertia and cost consciousness limit the effectiveness of cartographic communication. By making economical what until recently has been too costly, computers and digital cartographic data bases might enable map makers to pay greater heed to cartographic theory and the needs of map users.

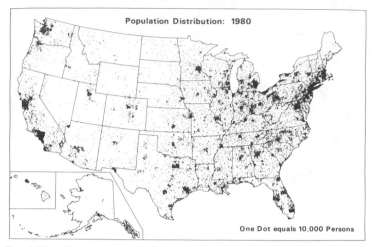

FIG. 4.22.   Computer-produced dot map.
Source: Courtesy Geography Division, U.S. Bureau of the Census.

Policy and Politics

Base mapping, remote sensing, and geographic data systems can
become the foci of intense public debate. Maps and aerial imagery
have advantages obvious to local planners, land developers, oil
companies, scientists, and the military, to name a few user groups.
Because public funding for remote sensing programs is limited,
competition can be keen among regions of the country, and between military and civilian applications. Even among scientists
priorities invite conflict: geologists and agriculturalists, for example, differ in their preferred spectral bands and time of day for
sensing. Map scale and sensor resolution are other points of dispute, and the form of geographic data—paper maps or computer
tapes—may provoke antagonism among competing users who
demand greater attention to their own needs and lower subsidies
for others.

Many of the questions likely to be asked reflect long-standing
themes in American political debate. Should there be any subsidy
at all, and should remotely sensed data be collected and owned by
government or by private investors? As with communications sat-

ellites, land resources satellites could be owned privately, with their imagery distributed through commercial channels, as is the news and other forms of information in most capitalist societies.[41] Is there sufficient "cream to be skimmed" by commercial entrepreneurs, and if so, how can the interests of those less able to pay be served? How much government regulation is appropriate? Can commercial ventures be coordinated? Is competition beneficial or detrimental? Can a monopoly or oligopoly be tolerated? Could government provide better service than the private sector? Must the military have its own systems? To what extent can military systems be shared with civilian users? How much ground detail can be tolerated without jeopardizing national security? Questions such as these have political, rather than scientific, answers.

With satellite imagery, policy concerns are international as well as internal. When not in a synchronous equatorial orbit, a satellite inevitably overflies other nations' territories and compromises their privacy and security. Third World countries, in particular, are highly apprehensive that unfriendly neighbors or multinational corporations will use remotely sensed imagery for military intelligence or resource exploitation.[42] Increased spectral sensitivity and resolution of 30 m [100 ft] or less hardly alleviate these fears, and the possibility of unregulated private ownership makes improved sensing and image processing systems seem even more ominous. Yet the use of Landsat imagery by less developed countries has demonstrated that developing nations can use satellite imagery to inventory their own resources, and thereby be compensated for potential losses of privacy.[43] A tactfully managed satellite sensing program can be an effective instrument of international cooperation and regional development.

CHAPTER 5

Decision Support Systems

Information about a nation's physical and social characteristics has many applications: in military operations and defense, in day-to-day administration of government operations, and in planning and evaluating a variety of development schemes from extracting mineral resources to providing social services. Geographic information such as the land-use and land-cover data discussed in the previous chapter is of little use without the efficient storage and ready retrieval provided by the digital computer. This chapter examines how computer technology is making map data more useful than when the principal medium was paper and the principal filing system was the wood or metal storage cabinet.

One of the modern manager's principal tools is the *decision support system,* abbreviated DSS and defined loosely as a computer-based system that provides quick support for the thought processes and decisions of one or more executives in the form of numbers, charts, and maps.[1] Key requirements for such a system include rapid response, easy operation, ready adaptation to change, and communication in familiar terms. In both government and private business, any program addressing regional or local differences needs accurate, up-to-date geographic information to guide policy formation. In an era of intense competition among firms and limited support for public works and social programs, intuition or guessing is no more suitable for describing places than for estimating costs. A *geographic information system* (GIS) is thus an important part of many *data base management systems* (DBMS). Computer graphics enhance the manager's ability to generate and experiment with spatial displays, and decision-makers are developing an increased appreciation of maps.

Policy analysis calls for maps different from those employed in highway planning or field geology. The scale usually is smaller because the user wants a graphic overview of an entire state, region, or nation. Indeed, a cartographic DSS treats maps not as sheets, as in topographic mapping, but as themes, or coverages. Large-scale base maps show a limited number of features, and decades are often required to complete a nation's coverage. Revision is a major task, usually pursued sheet by sheet rather than feature by feature, so that seldom are large areas covered with uniformly accurate information. Thematic maps, in contrast, are highly generalized, often to the extent that coastlines and political boundaries are little more than caricatures—cartographic cartoons, perhaps, but useful nonetheless. Counties, states, and even regions might each be associated with a single numerical value. An example is the map of the rate of population change between 1970 and 1980, shown in Figure 5.1. Severe distortions, as in the *area cartogram,* might be needed to give each symbol a visual weight in accord with the population it represents, as shown in Figure 5.2. Although its mission is quite different—to communicate an accurate generalized overview rather than geometrically accurate local detail—this type of cartographic display is no less a map than a large-scale topographic quadrangle sheet. Not all thematic maps are such seemingly simplistic schematic drawings, of course. Considerable thought and painstaking research underlie a good thematic map, particularly in the case of small-scale environmental displays not based on political or administrative areas. Perhaps the most enigmatic problem in cartography is the generalization to a much smaller scale of thematic data, such as land use, mapped at a larger scale.

Maps and Censuses

Censuses, the source of much data for modern small-scale maps, are about as ancient as government itself. As far back as 3800

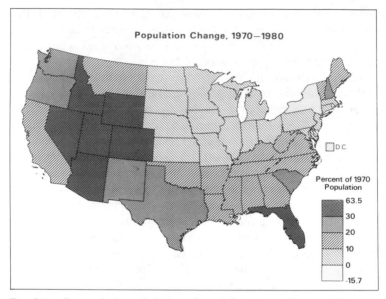

Fig. 5.1.   State-unit choropleth map of population change illustrates the generalization of boundaries and coastlines common to small-scale thematic maps.

B.C., Babylonian rulers counted the number of tax-paying sub-
jects. Censuses often were highly selective and the results not
widely published: the number of men available for military con-
scription, for instance, was a closely guarded secret. By 3000
B.C., the Chinese were counting population, and by 2300 B.C.,
they were preparing maps and written descriptions of not only
population but also agriculture, industry, and commerce.[2] Other
nations and cultures had censuses much earlier, but it was the
United States that, in 1790, initiated the oldest continuous periodic enumeration, with a decennial census required by the Con-
stitution for apportioning seats in the House of Representatives.
Nonetheless, modern demographic data, with age, sex, and other
descriptive information for all individuals in a household, were
not collected systematically by any country until the mid-nineteenth
century.

The mid-nineteenth century also marked the advent of the statistical map for providing a geographic overview of census tabu-

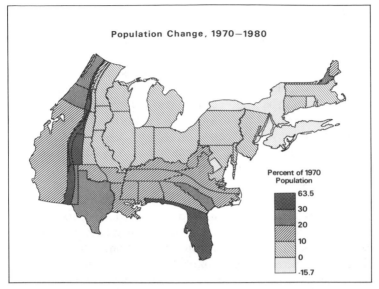

Population Change, 1970—1980

Percent of 1970
Population

| 63.5 |
| 30 |
| 20 |
| 10 |
| 0 |
| -15.7 |

FIG. 5.2.   Area cartogram deliberately distorts the map areas of data units to make bigger symbols for more important places and smaller symbols for less important places. In this case a state's area is proportional not to its land area, but to its mid-decade, 1975 population.

Source of cartogram base map: R. Eastman, B. Nelson, and G. Shields, *Isodemographic Map of North America, 1975–1976* (Kingston, Ontario: Cartographic Laboratory, Department of Geography, Queens University, 1978); used with permission.

lations. The development of statistics as a science and the establishment of government agencies to collect official statistics preceded statistical cartography.[3] The forerunner of Britain's Royal Statistical Society was founded in 1834, and by mid-century statistical bureaus had been set up in most European countries. Inspired by small-scale maps and atlases developed in Austria and Prussia, the United States census, which had published not a single map before 1854, began a vigorous program of statistical cartography in the early 1870s. In 1871, Francis A. Walker, Superintendent of the Ninth Census, loaned a collection of Prussian statistical maps to Daniel Coit Gilman, an eminent geographer and later President of The Johns Hopkins University. Highly impressed, Gilman en-

couraged Walker, who already had begun to map tabular geographic data. Inspired by an address by Gilman, the American
Geographical Society persuaded Congress to appropriate $25,000
for illustrating the 1870 census. In its summary reports, published
in 1872–1873, the Census included 12 maps on population distribution and characteristics, four maps on mortality, and six maps
on agriculture. The 1880 census provided even more maps, and
small-scale thematic maps have been important in the summary
reports of all subsequent censuses.

The major undertaking in preparing a small-scale map of an
economic or social theme for an extensive area is, of course,
collection of the data. Although surrogate information such as
multispectral scanner imagery is sometimes useful, land-cover
data alone provide a very limited picture of what occurs in homes,
factories, offices, and produce markets. Most comprehensive so-
cio-economic maps require census information, collected through
either a complete enumeration or a systematic sample survey designed to estimate an accurate value for each enumeration area
with sufficient population or economic activity to warrant consid-
eration. An alternative, particularly common in Scandinavian
countries, is the *registration system,* with data recorded on a continual basis for, as examples, births, automobile ownership, or
school enrollment. In the United States, registration systems are
limited largely to vital events (birth, death, marriage, and divorce)
and activities that are licensed or taxed.

If a census enumeration is to be complete and accurate, large-
scale maps are needed to record the boundaries of enumeration
districts, to relate these districts to existing political units for which
tabulations will be required, and to account for all streets and
roads along which the houses or businesses to be surveyed might
be located. Extensive field work is needed during the months
preceding the enumeration, especially in areas with new construction, apartment conversion, or building demolition. Local citizens
are often employed for several months in an effort to identify all
households. Aerial photography may be useful, especially in lesser
developed countries without a previous census. For metropolitan
areas the U.S. Bureau of the Census prepares urban area base
maps at 1:24,000, with all streets named and all blocks labeled.[4]

The Census Bureau also rents address lists from several large direct-mail advertising firms and cross-matches these lists in an effort to compile a complete but nonredundant list of households.[5] Local post offices and letter carriers cooperate by identifying all households receiving mail.

An intensive advertising campaign through television, radio, and the print media precedes the enumeration. Most people can be made to see the need for a national census and cooperate willingly. Most American households complete and return the self-administered mail-out and mail-back questionnaire used in large metropolitan areas since 1970, and enumerators need visit only a small proportion of all households.[6] Illegal aliens, illiterate or developmentally disabled persons, and those fearing contact with government tend not to respond. Questionnaires in Spanish are sent to Hispanic households. Returns are matched against the master list of households, and residents not responding are prodded by phone, mail, or personal visit.

Hand tabulation of summary statistics is error-prone and costly. Starting with the 1960 census, optical scanners similar to those used for test scoring have been used to convert pencil marks on the questionnaire to machine-readable responses stored on magnetic tape for sorting and tabulating by computer. These responses must be assigned to the smallest areal unit containing the household. In the United States the smallest geographic cell is the *block face,* essentially consisting of all households on the same side of the street between two successive cross streets. Two block faces back-to-back, together with any adjoining block faces, form a *block* (Figure 5.3). A carefully selected group of contiguous blocks with a population of about 4,000 is called a *census tract.* Tracts often reflect a compromise between describing cohesive urban neighborhoods and providing comparable units for studying the geographic patterns of change between censuses. Census tract boundaries often coincide with political boundaries so that counts for tracts can readily be aggregated upward into counts for villages, towns, cities, counties, metropolitan areas that comprise one or more whole counties, states, "divisions" such as New England that consist of several states, and regions such as the South that consist of two or three divisions.[7] Accurate assignment of

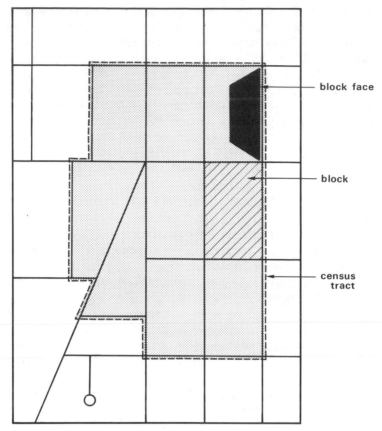

FIG. 5.3.   Block faces combine to form blocks, which, in turn, combine to form census tracts of approximately 4,000 residents.

residents at the block-face level is particularly important because the block is the fundamental unit for periodic legislative reapportionment schemes, and the block face is the link between the street address and the block.

DIME Files and Data Structures

A digital map called a DIME file relates households to the appropriate block face, block, and tract. DIME is an acronym for Dual

Independent Map Encoding. "Dual" refers to the two-fold role of the street segment in linking both the two intersections that *bound* it and the two blocks that *cobound* it. The DIME concept is more general in that the street segment is but one type of *edge*, or "1-cell," that can be cobounded by two adjoining blocks, or "2-cells," and the street intersection is but one type of *node*, or "0-cell," at which two or more edges meet. Other types of edge include portions of political boundaries, canals, railways, shorelines, and ridge crests. The DIME file consists of a list of nodes and their plane coordinates, and a list of edge records that specify, for an arbitrarily chosen direction along each edge, the identifying numbers of the beginning and ending nodes and the blocks to the left and right. This *topological* information allows the computer to link up edges around a block and check for errors in the file.[8] Address coding information, such as street name and type and the address ranges for the left and right sides of the street segment, is included in the edge records so that household addresses can be matched by computer with the appropriate block face, block, and tract (Figure 5.4).

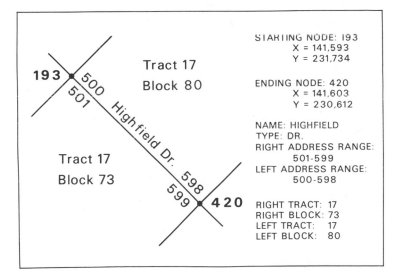

Tract 17
Block 80

Tract 17
Block 73

Highfield Dr.

193
500
501
598
599
420

STARTING NODE: 193
    X = 141,593
    Y = 231,734

ENDING NODE: 420
    X = 141,603
    Y = 230,612

NAME: HIGHFIELD
TYPE: DR.
RIGHT ADDRESS RANGE:
    501-599
LEFT ADDRESS RANGE:
    500-598

RIGHT TRACT: 17
RIGHT BLOCK: 73
LEFT TRACT: 17
LEFT BLOCK: 80

FIG. 5.4.   Contents of a DIME record for a street segment permit an address to be matched with its block and tract.

The DIME concept of edges that meet at nodes is the basis for the topological ordering of many data files in digital cartography. A notable example is the DIMECO file, developed by the U.S. Bureau of the Census to promote the computer-assisted preparation of statistical maps based on county-unit data from census tabulations and other sources.[9] Each edge is a shared, common boundary between two adjoining counties. In this case the edges, or *chains,* are represented not only by the coordinates of their beginning and ending nodes, but also by the coordinates of all intervening points needed to describe the character of the boundary segment at scales of 1:5,000,000 or smaller. Choropleth maps can be created by joining together the chains that bound each polygon representing a county and filling the interior with a shading appropriate to its data value (Figure 5.5). Chains between adjacent counties assigned to the same mapping category may be suppressed to enhance the visual impression of the larger category-regions. Because each common boundary is represented just once, this data structure eliminates the overlap or underlap likely if each county is recorded as a separate and independent polygon.[10]

Another data structure useful for mapping census data in a decision support system is the grid. Instead of aggregation to irregularly shaped political units varying widely in size and shape, household returns may be aggregated to the square cells of a uniform grid based upon a recognized plane coordinate system. Cell size must accommodate the scale of the intended maps and the spatial resolution of the mapped phenomena. If the cells are too large, the mapped pattern will appear blocky, and the grid may dilute significant contrasts between adjacent dissimilar areas as well as split comparatively homogeneous zones.

One of the more ambitious attempts at grid mapping was the aggregation of 1971 census data for Britain's 18 million households to 1-km square cells based on the National Grid system of plane coordinates. Each cell measured approximately 0.25 mm [0.01 in.] on a side on the 34 page-size, 1:4,000,000-scale nationwide maps of a four-color experimental atlas published in 1980.[11] More than half of the 152,440 grid cells contained either no data or too few households from which to compute reliable ratios and percentages. Although perhaps not cartographically

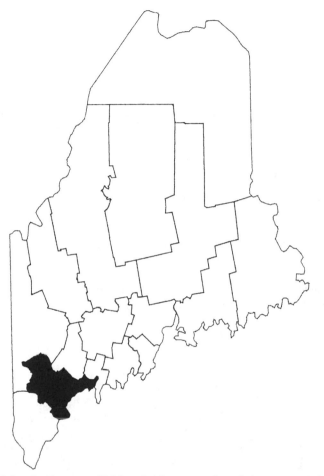

F_{IG.} 5.5.   Outline map of Maine showing county boundaries as represented in
the DIMECO file. Black area is Cumberland County.

elegant, these maps, with a majority of blank cells, were at least
more geographically valid than most small-scale maps of census
data.

Britain's grid maps are but one refinement of the much coarser,
less aesthetically pleasing black-and-white grid maps produced in
the late 1950s and 1960s on computer-driven line printers de-

signed principally for tabular printout. Among countries involved in the early development of these crude computer maps was Sweden; together with other Scandinavian nations and the Netherlands, Sweden has one of the most comprehensive programs for the collection of demographic data. Since the middle of the eighteenth century, Sweden has maintained a continuous registry of births, deaths, and marriages. It was also a leader in developing a variety of "geocoded" data bases important to national and regional planning. In addition to an interactive, conversational system for storing and processing census data, Sweden maintains an on-line cadastre, a computerized information system describing all public roads for the analysis of highway accidents and traffic flow, a nationwide digital elevation model, a digital land use inventory, a national digital map at 1:250,000, and several other regional geographic information systems.[12] Continuing as a leader in geographic data base management and computer-assisted mapping, Sweden demonstrates that techniques useful for census data can be applied to a variety of other statistical data collected on a continual and routine basis by the central government of a developed nation.

Since the 1950s, computer-assisted cartography has made substantial progress in efficiency, variety of symbols available, and aesthetics. The first computer-produced maps were produced on electric typewriters used as display devices by early electronic computers. The look of these maps was constrained by alphabetic and numeric print characters more tall than wide arranged by row and column (Figure 5.6). Grid cells could be blank but not solid black: since typed characters could not touch, the separation between adjoining characters and successive rows left light-toned gaps even when several different, superimposed characters were overprinted to blacken most of the cell interior. Printers producing as many as 2,000 lines per minute replaced the relatively slow electric typewriter, and geographers and others began to generate reams and reams of maps, often with neither careful planning nor thoughtful examination. Most line-printer maps were either choropleth displays, with shading symbols confined by the boundaries of political units or enumeration districts (Figure 5.6), or

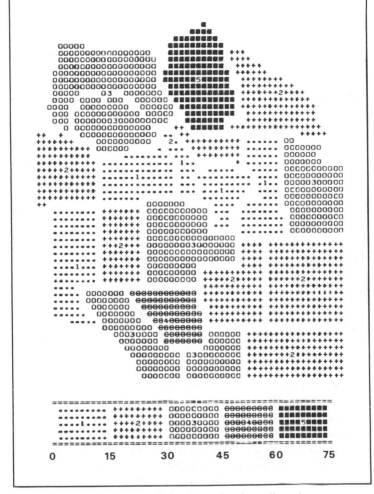

FIG. 5.6.   Example of a choropleth map produced on a line printer.

contour maps, with lines of equal surface elevation represented by chains of blank cells between shaded inter-contour regions (Figure 5.7).

Line-printer characters were usually spaced ten to the inch along the row so that the graphic resolution of the line printer was no finer than 2.5 mm [0.1 in.]. In the mid-1960s, the pen plotter, with a resolution of 0.25 mm [.01 in.] or finer, greatly increased the aesthetic quality of computer-produced maps. Two types of plotter evolved, drum and flatbed. Drum plotters move the pen along the X axis by rotating a drum with sprockets to guide the roll of paper perforated at its edges; this drum pulls the paper toward either the supply or take-up roller (Figure 5.8, upper). Flatbed plotters draw on a firmly anchored, flat sheet and move the pen in the X direction by shifting to the left or right a gantry with a sliding pen holder (Figure 5.8, lower). Both types of plotter draw in the Y direction by sliding the pen holder along a straight track, perpendicular to the X axis; this track is stationary on the drum plotter but mounted on the movable gantry of the flatbed plotter. Simultaneous movement in both the X and Y directions produces a diagonal line, and the pen can be raised at the end of one linear feature and moved in the "up" position to the beginning of the next. Resolution is often better than .025 mm [.001 in], and lines can now be drawn with neither the blockiness of the coarse line-printer grid nor the jerky, "stepped" diagonal lines of early pen plotters. A variety of line, point, and area shading symbols are possible, as is generally acceptable text for titles and place names (Figure 5.9).

Line printer and pen plotter require substantially different data structures. Line printers display maps row by row so that the data must be organized line by line beforehand in a *raster* data structure, named after the rake-like pattern of lines formed by the scan of the electron beam across the inner surface of a cathode ray tube. In contrast, pen plotters merely require that the coordinates of points describing a linear feature or the closed boundary of an area polygon be organized in the required plotting sequence. This organization of point data is called a *vector* data structure because each feature is represented by a sequence of vectors or directed

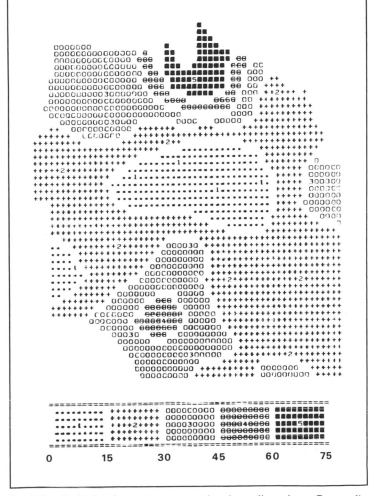

Fig. 5.7.   Example of a contour map produced on a line printer. Contour lines
are represented by the blank gaps between the shaded inter-contour areas.

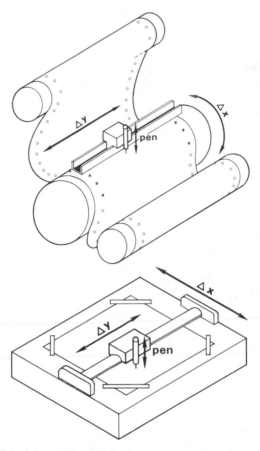

FIG. 5.8.   Principles of X and Y displacement produced by moving gantry of flatbed plotter (below) and rotating drum of a drum plotter (above).

line segments. Given this internal organization, though, the individual features may be plotted in any order. Although not required for graphic display on a pen plotter, vector data can benefit from the topological order of a DIME file. Many processes are best served by a somewhat redundant *hybrid* data structure, composed of both raster and vector representations.

Vector data commonly require less storage space than raster

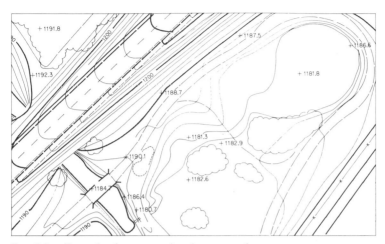

FIG. 5.9.   Example of a map produced on a pen plotter.
Source: Courtesy Synercom Technology, Inc.

data. Fine-resolution grid maps covering large areas can consume enormous amounts of storage, especially when the mapped distribution does not exhibit frequent, radical local variations. When computer memory is limited, modes of raster organization more efficient than the grid are needed for storing these *sparse matrices,* so called because relatively few of the cells represent a linear or point feature. One common solution is *run-length encoding,* whereby each scan line is divided into continuous groups of cells with the same attribute (Figure 5.10). Instead of requiring a piece of memory, or *word,* for every cell, each scan-line segment need be represented by only a pair of words—its attribute and number of cells—with successive pairs organized in a list according to the sequence of segments from left to right across the scan line. Compressing the data in this fashion can not only reduce storage requirements to a fourth or less of that required to store the full grid, but because fewer pieces of data are involved, compression also can decrease significantly the time and computational effort required to process the data.[13] Designers of decision support systems use data compression to cope with large amounts of geographic data and slow long-distance communications channels.

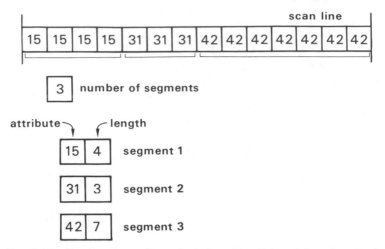

FIG. 5.10.   Run-length encoding reduces from 14 to 7 the number of words of memory required to described this sample scan line.

Displays, Pointers, and Interactive Maps

Both line printer and pen plotter are relatively slow display devices, unsuited to the interactive review and editing requirements of large, complex cartographic data bases. In contrast, the *cathode ray tube,* or CRT, provides the speed needed for the rapid visual inspection of different features, at different scales, for different parts of the mapped region.[14] An image is formed by a beam of electrons striking the phosphorescent coating inside a large vacuum tube (Figure 5.11). Beam intensity controls the brightness of the fluorescing phosphor. In a *raster-scan* CRT the electron beam sweeps across the screen in parallel horizontal lines, scanning from top to bottom in about one-thirtieth of a second. The screen must be refreshed from a special memory buffer at least 30 times a second—any less frequently and the screen will appear to flicker. A *directed-beam* CRT must also be refreshed, but its electron beam can trace any figure on the screen in vector mode. A *storage* CRT likewise uses a freely directed beam to plot vector figures but requires no image refresh. Its picture may be retained for an hour or longer and can be drawn slowly by a less

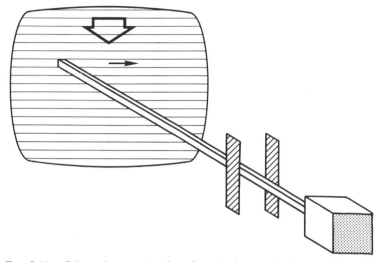

FIG. 5.11.   Schematic representation of a cathode ray tube. Beam of electrons (cathode ray) is directed at the phosphorescent inner surface of the screen. The phosophor fluoresces as it acts as an anode to absorb the electrons. Electromagnets direct the beam in a raster sweep along each successive scan line, back and forth from top to bottom down the screen. Beam intensity can be varied to provide a range of gray tones. Electron gun, electromagnets, and phosphor are enclosed within a vacuum tube to avoid interference from dust or moisture.

sophisticated, less expensive processor without the buffer memory required to support a refresh CRT. Unfortunately, the entire screen must be erased if any part of the drawing is to be deleted, and a couple of minutes may be required to display a map with many features or complicated symbols. Nonetheless, storage CRTs generally provide a sharper, finer-resolution image of line drawings than raster-scan, refresh CRTs.

Some raster-scan CRTs can display maps and other pictures in color.[15] Instead of the standard phosphor used for black-and-white pictures, the inside of the screen is coated with an array of phosphor dots that emit but one kind of light—blue, green, or red. Instead of a single electron beam, three beams sweep the screen simultaneously—one for the blue dots, one for the green, and one for the red. A metal mask with thousands of small, precisely cut and positioned holes is placed between the electron-beam generator and the phosphors so that each beam will strike only those dots producing its designated color. Thus where the screen is to be colored red, only the red beam will be intense. For an area to be colored yellow, both the green and red beams will be intense. Any color can be generated from these three additive primary colors.[16] White requires intense beams for all three colors, whereas black results when all three "beams" carry no energy.

Other rapid display devices are in use, and the CRT is, in fact, just one variety of *visual display unit* (VDU)—a generic term likely to be used more widely in future years. One example of another kind of VDU is the *plasma panel*. "Plasma" describes the thin layer of gas sealed between two parallel glass plates to provide an indefinitely permanent, stored image from which parts of the picture may be added or erased without affecting other parts.[17] When ionized by an electrical discharge, the gas emits light. The panel's raster image is formed on a 512 by 512 grid of thin conductors, typically with 512 rows printed on one inner surface and aligned precisely and perpendicularly to 512 columns on the opposite inner surface. Each of the 262,144 points (512 by 512) at which a row-wise conductor is close to a column-wise conductor is the center of a display cell. Barriers prevent the luminous krypton, neon, or xenon gas from drifting from one cell to another. A gas discharge can be fired at cell (54, 283), for example, by generating a brief flow of high voltage electrical current between the conductor in the 54th row and the conductor in the 283rd column. This minute gas discharge will emit light because stripping some of their electrons has excited the gas molecules to a higher energy state. Similarly, a negating charge to restore these electrons will extinguish the glow. Although the pixels are organized as a grid, they may be addressed in random order. The glow may last for up to an hour so that a storage buffer for these raster data may not be necessary. Simple conversion techniques that fill in the missing cells between points in a list of coordinates promote the rapid display of vector data. Larger, more detailed grids are possible. Although individual panels seem limited to a 1,024 by 1,024 raster, multiple panels could be arrayed to provide a much larger, even wall-size, display.

Optical rear-screen projection is the basis for another large-screen VDU, a four-color system with 2,048 by 2,048 addressable pixels projected from a 2.54 cm square [1 in. square] *liquid crystal light valve* onto a 122 cm square [48 in. square] screen.[18] The liquid crystal operates as a mask to either block or pass light from a source behind the projector (Figure 5.12). When the light is allowed to pass, a bright, enlarged image of the pixel appears on

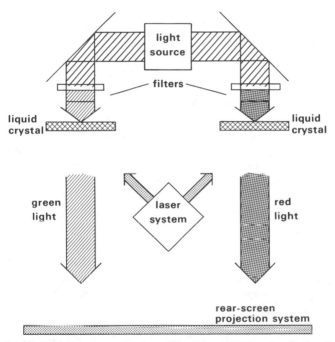

FIG. 5.12. Schematic representation of liquid crystal light valve display. Laser beam heats pixels on 2,048 by 2,048 grid of liquid crystal light valve to block emission from light source on other side of crystal. Filters yield separate beams of red and green light to be blocked wholly or partly at each corresponding pixel on separate liquid crystals. Each modified beam is then projected onto the rear of the screen, with the enlarged grids of red and green pixels in registration. A pixel receiving both red and green light will appear yellow.

Source: Robert C. Tsai, "High Data Density 4-color LCD System," *Information Display* May 1981: 3–6.

the screen. These enlarged pixels may be turned off or reduced in intensity by a scanning focus lens that guides a laser beam from pixel to pixel on the liquid crystal. When the laser heats a pixel intensely, the light valve blocks the light and the corresponding pixel on the screen is darkened. When the beam is less intense, the light valve passes light and the screen pixel is bright. Beam energy can be varied to produce different gray levels. Two liquid crystal light valves can be used, one for green and the other for red. This system generates a third color, yellow, where both red and green light are directed toward the same pixel on the projection screen. The third additive primary color, blue, could be added with a third liquid crystal to provide a full choice of intermediate hues. Like the monochromatic plasma panel, the liquid crystal panel forms an image of indefinite duration that may be erased as well as addressed both randomly and in raster mode.

Optical principles have been exploited further, to generate three-dimensional images. Synthetic holograms generated on photographic film by a computer-controlled laser beam can exploit the principles of light wave diffraction and coherent optics to provide an image that not only appears three-dimensional, but also may be made to rotate so that the observer can view it from a range of directions.[19] This ability is particularly useful for oblique surface maps, which often suppress otherwise useful information by deleting hidden lines to promote the effect of a solid body (Figure 5.13). Other approaches to three-dimensional maps include the display of stereo images, either side by side on a small screen so that one eye is focused on each image, or superimposed in, for example, red and green for anaglyph viewing through glasses with a red filter for one eye and a green filter for the other.[20] Further advances in display technology are likely to follow from a renewed exploration of such alternatives to the CRT as electroluminescent-layer displays, light-emitting-diode displays, gas-discharge displays, and liquid-crystal displays.[21] The CRT is but one of many electronic approaches to graphic display.

Perhaps as important as a rapid VDU for a decision support system is a convenient mechanism for selecting rectangular windows on the screen for enlargement or for indicating points or

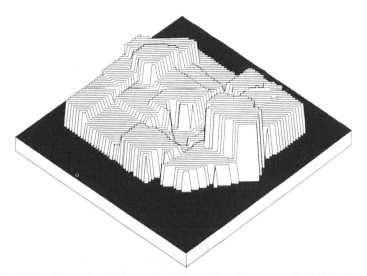

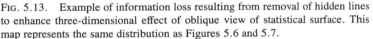

Fig. 5.13. Example of information loss resulting from removal of hidden lines to enhance three-dimensional effect of oblique view of statistical surface. This map represents the same distribution as Figures 5.6 and 5.7.

features about which further information is to be displayed or entered. A variety of pointing devices have been developed for interactive computer graphics. Among the earliest was the *light pen,* actually a light detector used with a raster scan CRT to record the time elapsed since the electron beam left an origin, for example, at the upper left of the screen.[22] Because the beam's horizontal and vertical rates of travel are known, this precisely measured time converts readily into a pair of down and across coordinates. Other pointing devices allow the user to move across the screen a target formed by a pair of short, intersecting, perpendicular lines. The system knows the instantaneous position of the target. Movement is controlled by a *joystick,* a lever that may be moved in two dimensions, or a *tracker ball,* mounted to rotate in two dimensions about a stationary center. The hand steers the target around the screen by "rolling" the tracker ball in the intended direction. Another kind of pointer is the *digital tablet,* a flat surface on which the position of a stylus or the intersection of cross hairs in a cursor is sensed by one of a variety of acoustic, magnetic, or electric sensors.[23]

The digital tablet may also be used with a *menu* of system commands to "talk" to the system. A different operation is printed in each box of a grid placed on the tablet. The analyst positions the stylus or crosshairs in the cell of the intended instruction, the tablet records the pointer's X,Y coordinates, and the system then determines which box was selected and performs the desired operation. Because the manager need not memorize the list of system commands and their exact spellings, and because pointing is faster than typing, a digital tablet menu is usually more efficient than a keyboard for entering instructions. Still more advanced, though, are voice input systems that can recognize speech and accept feature codes and commands more rapidly than possible with a keyboard or digital tablet.[24]

The variety of hardware for processing graphic data is great. Manually directed digitizers record X,Y coordinates for "capturing" vector data, and high-speed optical scanners record reflected light from a drawing mounted on a rotating drum to capture raster data.[25] In addition to the interactive "soft copy" display and pointing devices discussed here, "hard copy" printers and plotters are now more versatile than the comparatively crude but nearly ubiquitous line printers and pen plotters discussed earlier. (Chapter Six discusses more fully these developments, including so-called film writers for producing either microfilm graphics or the much larger final negatives from which press plates may be made for color lithographic printing.) The ability of hardware manufacturers to mass produce graphic display units for artists, designers, managers, and cartographers has been demonstrated, as has the ability of software developers to provide efficient, "user friendly" computer programs for many different applications. Although many innovations await development and many existing systems require extensive modification—if not complete replacement—it would appear that most of the significant technical problems have been solved or nearly so. In many ways, good management is now more critical than better technology to the successful implementation of computer graphics.

National Atlases and Data Banks

For cartographic applications of computer graphics, critical challenges involve the collection, standardization, and organization of huge amounts of data, at a variety of scales and for many different types of feature. Standardization is particularly important for decision makers, who are better able to detect a correspondence among geographic distributions if a consistent map projection, a few standard map scales, and a common set of symbols are used. Although unique, innovative projections can sometimes aid analysis and communication, such variety can be distracting. Fluctuating scales, different symbols representing the same phenomenon, and similar symbols representing different phenomena also confound the visual cross-correlation of two or more maps. The ability to retrieve, display, and compare all relevant geographic distributions without graphic distractions should promote the serendipitous discovery that often accompanies "putting two and two together." A carefully standardized geographic data base should encourage this so-called bisociative thought, recognized by writer Arthur Koestler as the cognitive basis for many significant discoveries in science and art.[26]

Data standards must promote accuracy as well as uniformity. The information in a data base should reflect the most advanced thinking and measurement skills while acknowledging legitimate scientific controversy. The advice and experience of not one but many experts should be sought in deciding what mappable data to collect and how to collect them, and in evaluating these compiled data and their cartographic representations. If limited time and funds dictate an incomplete or highly generalized interim map, peer review might not only yield useful refinements but also guarantee the inclusion of appropriate caveats.

The organization and support for this kind of comprehensive geographic data base traditionally has come from government, and the product is called a national atlas. A comparatively new phenomenon, less than a century old, the national atlas is not much more recent than the modern, demographically detailed census, upon which it is in part based. Finland published the first national atlas, in 1899, and Canada followed closely with the second, in 1906.[27] In 1960, K. A. Salichtchev, a Soviet professor chairing a committee on national atlases for the International Geographical Union, counted 26 published national atlases.[28] By 1979, according to F. J. Ormeling, Sr., President of the International Cartographic Association, the count had doubled, to between 50 and 60 atlases.[29] The United States was a relative latecomer, not publishing its national atlas until 1970.[30] Canada, in contrast, produced a second edition in 1915, a third in 1958, and a fourth in 1973. A fifth edition was planned for the 1980s.[31]

This rapid growth in national atlases reflects in large measure a concern for national identity and the development of natural resources on the part of Third World nations, newly independent former colonies in many cases. Atlases produced by lesser developed nations commonly are small, graphically simplistic, and far from comprehensive attempts to inventory mineral and other physical resources. Not surprisingly, they often fall far short of the high standards for design, production, content, and detail set by the trilingual *Atlas der Schweiz,* published as 96 separate sheets over the 14-year period 1965–1978. The Swiss national atlas, proposed as early as 1934, is the result of decades of planning.[32]

Canada's program illustrates the benefits of a national atlas to its citizens and their government. Although benefit-cost analysis is inherently imprecise, budget analysts in the Department of Energy, Mines and Resources were able to estimate a better than 2 to 1 benefit-cost ratio for the 4th edition of *The National Atlas of Canada*.[33] This estimate was based on extensive interviews with government officials and other users. The overall time savings to persons in need of geographic information was judged to be 60 times greater than the aggregate time required to compile and publish the atlas. A highly cautious calculation that assigned the user's time a value only one-tenth that of the atlas producer's time yielded the conservative yet significant 2-to-1 ratio.[34] This estimate includes atlas users ranging from curious school children to federal bureaucrats, scientific researchers, and private investment analysts.

Additional benefits accrue from the ready availability in Ottawa of National Atlas compilation materials, as well as the small staff of full-time research geographers responsible for revision and research. An important but easily overlooked adjunct to the printed maps, these information resources have been of use to Parliament and a variety of government ministries in such tasks as planning pipeline routes and evaluating economic potential. National Atlas staff also have helped publishers of educational materials, local boards developing tourist literature, private-sector publishers of specialized and general interest atlases, and scientists in a variety of disciplines. Most citizens benefit ultimately through reduced costs to private-sector users, more accurate and abundant information about their country, and more informed decisions by federal and provincial administrators and legislators.

The fullest benefit from investment in a national atlas program hinges on the widespread dissemination of timely, accurate information. Timely publication, though, conflicts with the common notion of the atlas as a book of maps. Indeed, the more advanced national atlas programs have for years sold separate map sheets as a means of avoiding not only the delay of waiting until all of the maps are drafted and printed, but also the artificial closure of having to wait another ten or more years for a revised bound edition. To support their concept of the atlas as a collection rather than a single bound volume, Switzerland and Canada have provided a large, stiff, flat box. Map sheets can be added as well as removed for easy comparison. The United States, in developing plans for a second edition of its national atlas, had decided upon a screw-post binder, to which sheets could be added with little difficulty.

For traditional booksellers and bookbuyers, though, the bound atlas is still an effective vehicle for marketing. A cooperative agreement between the Canadian government and Macmillan of Canada demonstrates how private publishers can be useful in producing and marketing a more traditional, bound collection of maps. Macmillan produced English and French bound editions of the fourth edition of the *National Atlas of Canada* and sold them in bookshops throughout Canada.[35] Moreover, Reader's Digest of Canada photographically reduced and modified some of the artwork for inclusion in its own bound atlas.[36]

Whether bound or boxed, two books or collections may be

more appropriate than one. A natural separation exists between intermediate-scale reference maps, requiring numerous sheets to cover the entire country, and small-scale thematic maps covering the nation on a single sheet or less. Canada, by providing two different yet complementary national atlases, can focus its revision efforts with greater efficiency as well as publish a more timely product. The *Canada Gazetteer Atlas,* published in 1980 with a comprehensive index of populated places and other named landscape features, will be separate from the solely thematic fifth edition of the *Atlas of Canada,* to be published in the late 1980s. A variety of subscription and purchase options are to be available, including folios containing all sheets related to major themes such as climate.[37] The Canadian program demonstrates perhaps the most flexible dissemination to date of national atlas information.

Modern telecommunications and data base technology will radically alter the organization and operation of national atlas programs. The paper atlas most likely will become secondary to the digital cartographic data base from which it will be derived—it could disappear entirely as a government publication, although its information might, for a time at least, be disseminated in print form by private publishers. In contrast, the ease with which digital data can be transmitted by wire or airwaves and stored, retrieved, manipulated, and displayed on a microprocessor in home or office suggests more timely analyses, a far greater level of use, a more informed citizenry, and a more responsive government. If managed effectively, the modern electronic national atlas should have a most impressive benefit-cost ratio.

Information policy will become a significant political issue. If the data base is properly designed and administered, taxpayers will demand ready if not cost-free, access to the system they support. Government, concerned not atypically with the accuracy and possible sensitivity of the data, might seek to suppress some information because of political repercussions or embarrassment. The roles and obligations of the public and private sectors will also change and require negotiation.

A key position in federal government should be that of National Data Base Administrator, who would supervise the National Atlas and also coordinate the collection and distribution of many nongeographic data. This Cabinet-level official would respond to free-dom-of-information requests from the public and the press as well as review, and possibly censor, the contributions of private organizations as well as public agencies. An elaborate security and peer review process would be needed not only to maintain appropriate confidentiality but also to avoid the consequences of poor data entering the system.[38] Data must be adequately documented, with a description of collection procedures and an assessment of limitations. Untrained users must be made aware of other possible definitions and measurements and of the reliability of samples. Ready availability does not guarantee quality. If poorly controlled, an easily accessible data base may do more harm than slower, more traditional mapping methods.

Coordination with state and local agencies will also be important for such a system would probably accommodate national overview data as well as detailed local maps. Information would need to flow efficiently in both directions, to promote the nation-wide collection of useful data and, once collected, to guarantee their fullest use. A program of regional atlases linked to a national atlas would extend a nation's consistently high standards of map design and geographic content to the special needs and interests of subnational units. Digital computing and telecommunications networks add yet another dimension—the need for efficient, widely recognized standards for data exchange.[39] As state cartographers must be able to talk with federal cartographers, state computer data bases must be able to communicate with federal computer data bases.

Turnkey Systems and Mapping Software

The national atlas of the future—and indeed world, regional, local, and thematic atlases as well—will be a data base designed for use with a computer graphics system. Instead of turning pages in a book of maps or even removing sheets from a boxed collection, the atlas user will manipulate maps on a microprocessor and interact with the displays before her on a VDU. If she wants more detail, she will increase the scale through the zoom command, progressively enlarging the territory near the center. If she wants

a more generalized choropleth map, she will change the classification limits and employ fewer categories. If she wants to compare two distributions, she might display two maps side by side
on a split screen or generate a new map, with symbols chosen to
show areas of similarity and disparity. If she wants to know what
distributions are similar to a map of particular interest, she can
ask the system to select, for example, the ten most similar distributions, which she may then preview. If she needs to prepare a
map to illustrate a report, she can add or delete features, change
symbols, and label important features. The resulting fine-tuned
map might even be transmitted to her company headquarters or
publisher through a telecommunications network—the same network that provided her with current data. Her professional colleagues, customers, or interested general readers might view her
geographic analysis on their own computer graphics systems, also
connected to a telecommunications network, and store the result
for future display on a small magnetic disc. Maps of specific
interest can be located more rapidly than with a paper atlas, and
the user can work more directly with the information. Unlike with
print publication, the data need not be confined by symbols and
layout cast on paper in dried ink.

Interactive map use of this type has three prerequisites: a well-
designed, flexible micro- or minicomputer; efficient computer
programs to manipulate and display the data; and a comprehensive, accessible geographic data base. The computer system, an
integrated group of electronic machines, is considered *hardware,*
in contrast to the more transportable, less tangible programs and
data, collectively called *software.* All three components must, of
course, be compatible, yet each might be produced by a different
firm. In the early 1970s, for instance, it was not uncommon for a
university computer center to process government census data on
an IBM computer using programs from DUALabs, a not-for-profit
software and consulting firm that specialized in assisting clients
with analyzing the highly detailed computer-tape tabulations of
the 1970 Census of Population.[40] It is common now, though, for
a firm to promote compatibility by providing at least two of these
three ingredients. IBM, for example, leases programs to its clients,
and DUALabs even sold machine-readable census data reformat-

ted for greater efficiency. The U.S. Geological Survey provides two versions of its digital 1:2,000,000-scale regional reference maps, developed for the National Atlas: one version has DIME-like topological properties, whereas the other is designed for use with the CAM (Cartographic Automatic Mapping) system for plotting outline maps according to various map projections.[41] The Harvard Laboratory for Computer Graphics and Spatial Analysis, which developed a variety of programs for statistical mapping, promoted its products with compatible, carefully edited versions of the DIMECO file, the Central Intelligence Agency's World Data Bank-I file of worldwide international boundaries, and the Census Bureau's files of metropolitan area census tract boundaries, among others.[42] The informed user is greatly concerned with machine-program-data compatibility, and the efforts of a producer of one component to assure its harmonious operation with the other constituents of a decision support system not only enhances the sales potential of the principal commodity but also generates substantial additional revenue from the secondary product.

Perhaps the single most significant breakthrough in the quest for compatibility is the *turnkey system,* so-called because the user need only "turn the key" on a newly delivered system to have a fully functioning graphics system. Implementing a sophisticated graphics workshop is not quite so simple for a business or government agency that must also collect its own data as well as anticipate future applications. However, many inexpensive but reliable personal computer systems do indeed offer almost instant turnkey response to the computer hobbyist or home-based consultant.[43] Despite needs to have additional skilled personnel, retrain existing staff, or reorganize its managerial structure, the large user also benefits greatly from a well-conceived turnkey system. Large- and medium-size systems, after all, require the compatibility not only of hardware and software but also of individual hardware components.

Component compatibility is especially important in cartographic systems because of the variety of so-called *peripheral* equipment employed in the capture and display of geographic data. The central computer must often be linked to a pen plotter, digitizer, scanner, hard-copy graphics printer, VDU, magnetic disc storage unit, and tape drive—all produced by different manufacturers and all requiring an *interface* to the "main frame" or another peripheral. In the 1960s and early 1970s, computer centers experienced enormous frustration with, for example, pen plotters that would not accept instructions from the computer on which mapping programs were to have been executed.[44] The turnkey vendor selects components for compatibility as well as efficiency, acquires considerable skill in designing complete systems and in advising clients on needs assessment and planning, chooses freely among the best peripherals on the market, influences the design decisions of hardware and software manufacturers, and can deliver a fully tested and guaranteed system of reliably interfaced hardware and harmonious software.

Many different systems have been developed with graphic processing as a primary or important secondary goal. The larger systems usually support other uses such as scientific computing or business data processing. Operation may be either interactive or *batch,* that is, with program runs added to a queue to be processed later, when time permits and with no possibility for additional direction from the user. The computer is designed to support either a single user at a time or multiple users, with the main memory partitioned to accommodate several programs simultaneously, each of which receives in rapid sequence for a brief moment the attention of the computer's central processor. Large interactive systems usually employ this time-share concept, which may also allow several batch programs to share the system simultaneously with a number of interactive users. Because many interactive users communicating with a computer through terminals are at any given moment merely pondering their next step or waiting for a result to be printed, a large time-share system commonly can support 40 or more terminals at once.

At the other end of the scale of computing power and sophistication is the *stand-alone* personal computer, small, completely self-contained, designed for the exclusive operation by a single user, and costing little more than a couple of high-quality color television sets.[45] Its small memory limits the complexity of the programs that might be run, and all programs must either be written by the user or acquired directly from a friend or commercial outlet. Many of these otherwise isolated microprocessors are so-called *smart terminals,* able to operate independently or as a time-share terminal linked by a telecommunications network to a larger computer.[46] The larger system that provides access to sophisticated programs and data might be used to develop and transmit a small data set for closer study later on the home or office system operating in stand-alone mode.

Special features that distinguish graphics microprocessors from other systems include not only the fine-resolution color VDU, pointing devices, and the pen plotter, but also built-in graphics processing functions that eliminate having to write one's own program code in order, for example, to plot circles or generate dashed lines.[47] Graphics routines available on permanent memory chips or plug-in memory modules called ROMs, for Read-Only Memory, allow the user to shade area polygons for choropleth maps, remove hidden lines on oblique views of three-dimensional statistical surfaces, and clip parts of features lying outside selected rectangular windows.[48] Cartographic users of interactive systems might eventually have no need whatever to write any part of their instructions in such conventional programming languages as BASIC, FORTRAN, or PASCAL—a set of carefully conceived mapmaking commands might provide all standard solutions to the cartographic problems of projection, generalization, symbolization, and map aesthetics.[49]

The development and distribution of good mapping software is essential to effective, honest mapmaking. Institutes, firms, agencies, and individuals that provide the computer programs with which hundreds of users make maps will exert an enormous influence on cartographic design and content.[50] Software transfer makes possible the dissemination of new, useful, highly effective techniques, yet it can promote with equal ease the perpetuation of untested traditions as well as the introduction of visually seductive but generally dysfunctional innovations. Program developers should pay as careful attention to the user's understanding of their products' limitations and range of solutions as they do to writing program code that will yield consistent, predictable results.

One particularly obvious type of potential abuse is the default value or option substituted by the program when the user chooses to neither supply a parameter nor select from among several available procedures. Providing default values can be a disservice if, for example, the untrained user is likely to assume blithely that a choropleth map should have five categories that split equally the range between the minimum and maximum data values.[51] There is much to be said for compelling the user to make all such decisions rather than relying on someone else's idea of what might be appropriate for the average map. Rather than encourage the user to abdicate responsibility, the program should provide information to help him make an informed decision for the data and application at hand.[52] Cartography as a profession should be concerned with the inventory and comparative evaluation of mapping software, with the clear and comprehensive documentation of program goals and system operation, and with the proper training of program users.

Standards, Cooperation, and Shared Benefits

Information systems designed to support the decisions of managers require the support of widely held standards of accuracy, completeness, reliability, and compatibility. Users, who often tender to computers and data bases a trust rivaled only by that accorded the printed word, need assurance that the data are valid. Review panels and an editorial process similar in principle to those of scientific journals are needed, as is clear documentation, available in both abstracted and detailed forms. In addition to the imprimatur and provenance, the data should also have an internal organization conforming to a recognized standard. Just as electric shavers would be far more useful to international travelers if voltage, current frequency, and receptacles were uniform worldwide, data exchange likewise is more efficient if magnetic tapes and ROMs could be "plugged in" anywhere, perhaps with the aid of one of a limited number of standard interfaces.

Computer programs should also conform to recognized standards. Users require not only complete, lucid documentation and training manuals, but also some guarantee that the program has been tested to produce valid results for all combinations of options and parameters. Compatibility of program and data is important as well, and the establishment of recognized data standards will eliminate guesswork and arbitrary decisions by program authors. Linkages among programs are also important. A major breakthrough in the early 1980s was the SAS/GRAPH extension, which made mapping and other graphics routines readily available to users of SAS (Statistical Analysis System), a widely distributed software system for statistical analysis.[53] Program-program and program-data synchronization will become still more important as computer networks and personal computers make many more maps and many more mapmakers.

Standardization can be both good and bad. Although they promote the exchange of software and the integration of data, programs, and hardware, rigid standards might also stifle worthwhile innovation. Ideas for more efficient computers, peripherals, and algorithms might be excluded from a marketplace wedded to established standards and familiar processes as strongly entrenched as the tradition and inertia that bind the automobile industry to the internal combustion engine. The hasty, politically expedient imposition of standards might greatly lower the benefits of coordination. Goodwill and good thinking must motivate and guide the evolution of standards for cartographic data, mapping software, and graphic hardware.

CHAPTER 6

Map Publishing
and the Digital Map

The success of a map often depends upon how effectively it is published. And to publish a map, the mapmaker must have reason to assume the firm can at least recover production costs. In the private sector cost recovery is through sales, which depend upon public awareness of the product and customer satisfaction. In the public sector costs must be matched by legislative appropriation, departmental budget allocation, or interagency transfer of funds, and thus upon the mapping agency's ability to satisfy the perceived needs of citizens, civil engineers, or the military. Both public agencies and private firms must be concerned with marketing and distribution. Hence, the ultimate utility of any map depends as much on its reproduction and distribution as upon its design and symbolization. If academic cartographers have tended to neglect these facets of map communication, it is because map printing and shipment have been comparatively straightforward.

But have conventional printing and distribution promoted a truly effective and full use of maps? That cartographic reproduction technology and marketing strategies might have performed at levels comparable to other information media merely begs the question. Map information often is timely, and if several years elapse before a map reaches the user, its value is diminished. Moreover, if the initial reproduction and distribution cycle is slow, the revision cycle, which should increase utility by correcting errors and adding new features, probably is retarded as well. Indeed, when the technology of traditional map publishing requires cyclic revision rather than continual updating as errors are detected and changes occur, the accuracy of map information surely suffers.

Perhaps the most significant contribution of modern electronics to mapping will be the increased speed at which maps and map data can be distributed to the user. Like book publishing, map publishing involves a traditional manufacturing process in which authors and editors interact with various craftsmen as the manuscript is slowly but painstakingly transformed into the finished product. Rarely does a topographic map emerge from the printing press in less than a year after the surveyor and stereo compiler have indicated where its lines should be drawn. Yet the advances in printing and electronic data storage and transmission explored in this chapter can reduce dramatically the time lost in map finishing, eliminate much of the cost of warehousing and shipping associated with paper maps, and enhance both the accuracy and utility of map information.

Mapping technology is becoming ever more complex. The variety of physical science terms and principles sampled in this chapter attests to the almost frantic pace of development and to the many designs now or previously explored. Most significant, though, are the institutional advances that seem inevitable at a time when anyone with a home computer and an inexpensive graphic display device might use a digital map to become her or his own cartographer. Clearly the role of the map publisher will change greatly. This chapter begins with an examination of printing, the first great transition in map publishing.

Printing, Photography, and the Spread of the Map

Whatever changes affect maps in the next few decades should be viewed as yet another revolution in the technology of cartographic image transfer. Two earlier innovations that radically altered mapmaking and map use are printing and photography. Before printing, maps were copied by hand, by craftsmen called copyists. Few people owned maps, and few maps existed. A map user could seldom be certain that a mark on a copied map was placed there deliberately by the original cartographer—it might well reflect merely an intellectual quirk or nervous tremor of the copyist's hand.[1] In contrast, printing provided many exact duplicates and lowered production cost per unit, whereas photography introduced further economies and permitted the printing of any image that could be drawn. Printing vastly expanded the number of accurate maps, and photography greatly increased the number of persons who could prepare for production visually satisfactory, convincing maps.

These technological revolutions were by no means sudden, in the sense of political upheavals. Johann Gutenberg, a German goldsmith in Mainz, is given credit for developing movable type some years before 1457, the publication date of the Mainz Psalter (the first recorded application of movable type). However, Chinese printers using a different concept, block printing, had been reproducing images at least a thousand years earlier.[2] The first surviving map printed in Europe, a simple relief woodcut included in a reprinted seventh-century dictionary as a textual illustration, was not produced until 1472.[3] The next five centuries produced numerous incremental improvements in paper, ink, presses, engraving, and plate materials. At least four major printing processes were employed in map printing: during the sixteenth century, relief printing from woodcut blocks was largely replaced by intaglio engraving and copperplate printing, to be succeeded by letterpress relief printing and lithographic (literally, stone writing) planar printing toward the end of the nineteenth century. Lithography almost completely displaced letterpress during the 1960s.

Photography evolved in a similarly transitory manner, but with seemingly more rapid progress over a comparatively short time. Although relevant experiments with lenses and photosensitive chemicals preceded its acknowledged founders by several centuries, photography did not become a practicable technology until Joseph Nicephore Niepce made the first permanent camera image in 1826 and his later partner, Louis Jacques Mande Daguerre, used a salt solution in 1835 to fix an image more effectively than ever before.[4] Numerous independent inventors and corporations have further advanced photographic technology to the stage of the instant-print camera, color film, and the Xerox copier.

Photography has had a profound effect upon printing, particularly through the replacement of cast, movable, metal "hot" type by photoset "cold" type, whereby images of alphanumeric characters are transferred photographically from a master positive or negative. Equally significant have been photographic methods for printing halftone pictures and transferring an image from a negative to the press plate. Advances in photographic image transfer, in fact, underlie the dominance today of offset lithographic printing, a technique that has little resemblance to the "stone printing" of yesterday. Most maps are now printed by photo-offset lithography.

How does modern lithographic printing work, and what does it have to do with stones? The answers lie in the familiar facts that oil and water do not mix and that some rocks, commonly limestone, can be ground and polished to a smooth, flat, durable surface. If no thicker than about 10–15 cm [4–6 in.], this flat surface can be used in a press (Figure 6.1). The image is drawn on the

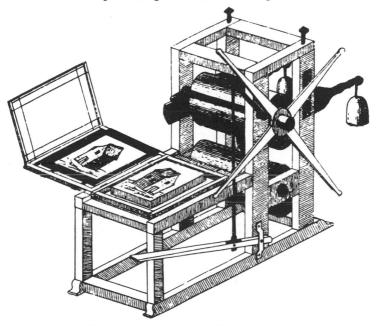

FIG. 6.1. Early lithographic printing press with stone plate.

Source: Alois Senefelder, *The Invention of Lithography* (New York: The Fuchs and Lang Manufacturing Company, 1911), p. 222.

stone with a grease pencil or crayon. Using a sponge, the printer then coats the surface with an etch mixture of water, acid, and gum arabic. The greasy image area on the stone repels this solution, which adheres to the surface in the non-image area. The weak nitric acid, which does not actually corrode the stone, enables the etch to spread and set firmly in the non-image area. Ink is then applied to the surface with a hand roller. Made from pigment and linseed oil, among other ingredients, the ink adheres to the waxy image drawn with the crayon but is rejected by the watery solution in the non-image area. If a sheet of paper is placed on the stone, with a flat plate placed on top and pressure applied, the image is transferred to the paper. This process is still used by skilled printmakers to reproduce the delicate details of quality stone-lithograph art prints.[5]

Lithographic printing requires not stone, just a smooth surface. Alois Senefelder, the Munich printer who invented lithography in 1796, realized by 1818 that metal plates could be more efficient. He had printed maps with stone lithography as early as 1808. By 1825, although widespread in map printing, lithography was still subordinate to engraved copperplates, with the line image cut into and below the surface to receive ink for intaglio printing. Nonetheless, stone was cumbersome, and lithographic images could more conveniently be drawn on zinc plates. But more than metal plates were needed. A lithographic figure, drawn with a crayon on special paper with a water-soluble coating, could be transferred to zinc or stone. The mapmaker could thus draw a direct, right-reading image, as it would appear on the print, instead of the reverse, wrong-reading image engraved on stone for the one-step transfer to paper.[6]

Photographic image transfer, vital to modern offset lithographic printing, did not reach its full potential until about 1940. In 1850, the basic principles of photochemical reproduction were known, but the artificial illumination required for most indoor work was not available until the carbon arc lamp was introduced in 1880. Images could be transferred photographically from negatives on glass plates, but these negatives were fragile and difficult to handle. The development during the 1940s of dimensionally stable plastics to serve as a base for photographic emulsions provided a significant missing link in contemporary map reproduction. Also important was the retirement of skilled cooperplate engravers, the impetus from aerial photography and photogrammetry to accelerate map finishing, and the greater efficiency of the rotary offset lithographic press, developed early in the twentieth century.[7]

The offset concept reduces wear on the plate and promotes printing thousands of impressions per hour. A right-reading metal printing plate is mounted on a rotating plate cylinder, to which first water and then ink are applied (Figure 6.2). As with stone lithography, the waxy image elements repel the water but accept the greasy ink, whereas the wetted non-image areas repel the ink. Rollers assure an even application of water and ink. The plate cylinder then transfers the inked image to the rubber-coated sur-

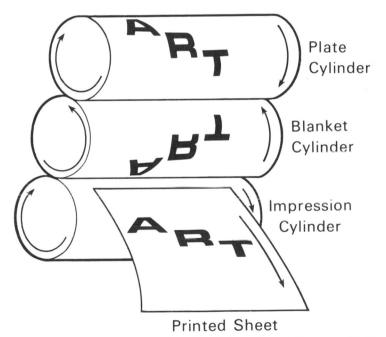

Plate
Cylinder

Blanket
Cylinder

Impression
Cylinder

Printed Sheet

FIG. 6.2.   Generalized model of an offset lithographic printing press. Right-reading image on plate is offset onto blanket cylinder as a wrong-reading image. A second offset produces a right-reading printed image as a sheet of paper is pressed between the blanket cylinder and the impression cylinder.

face of the blanket cylinder, on which the image is wrong-reading, or reversed like a mirror image. Rotating a bit further, the blanket cylinder again "offsets" the image, this time onto a sheet of paper that is pressed between the blanket cylinder and the impression cylinder. As intended, the image once again is right-reading.[8]

Before printing is attempted, the right-reading, positive image on the thin aluminum plate is transferred photographically from a sheet of film with a wrong-reading, negative image. The plate is first covered with a photosensitive emulsion and held in direct contact with the image side of the negative, as shown in Figure 6.3. Strong light passing through the clear image areas of the negative exposes the emulsion below. The opaque parts of the negative block light from exposing the non-image areas on the plate emulsion. After passage through an automatic processor with chemicals that remove the emulsion in the unexposed, non-image areas and develop and stabilize a waxy coating in the exposed, image areas, the right-reading plate is ready to be mounted on the press.[9]

The method of reproduction affects greatly the appearance of the map. Woodcuts, in which the non-image areas are carved out so that the image elements are raised to receive ink, were not

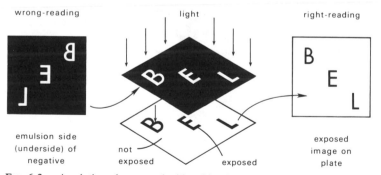

FIG. 6.3.   A printing plate coated with a thin photosensitive emulsion is placed in contact with the emulsion side (right) of a photographic negative and exposed to light. The wrong-reading negative blocks light in the non-image areas and transmits light through the image areas (center). A right-reading, positive image (left) forms on the plate as photographic processing removes the emulsion for the non-image areas and develops and fixes a hard, waxy coating in the exposed, image areas.

easily lettered and could not include area symbols with fine tones and other delicate details. With copperplates for intaglio printing, fine details could be developed as fine lines. Yet engraving was slow, with experienced engravers averaging only several square centimeters per day.[10] Wax engraving, a technique developed in America in the late nineteenth century, allowed the integration of fine linework and type.[11] A variety of engraving tools was used, including stamps for rules, area tones, point symbols, and letters. Wax-engraved maps typically were highly stylized, with numerous labels and other mechanically stiff symbols (Figure 6.4). Lithography, which seldom achieved the fine detail of copperplate engraving, eventually proved more versatile in its comparatively rapid generation of area shading tones.

Photographic lithography, which allowed the printing of anything that could be drawn, made cartographic drafting too easy, in a sense, and filled professional journals, books, government reports, and commercial street maps with the sloppy artwork of poorly trained non-craftsmen.[12] Moreover, lithography greatly changed the relationship between mapmaker and printer. From 1472, the date of the first recorded printed map, through about 1570, there was considerable experimentation as cartographers and others explored the printer's evolving abilities and printers adapted to the rising expectations of their clients. For the next two hundred years, copperplate engraving and intaglio printing were well established and the cartographer-printer relationship was comparatively formal. Numerous innovations, particularly lithography and photography, as well as the technological upheaval of the Industrial Revolution made the period from about 1780 through 1940 one of instability and change. Increased professionalism and sophistication among map producers together with the evolution of effective, widely accepted methods for the efficient preparation and printing of maps has led cartographic historian Arthur Robinson to label the interval since 1940 the Period of Technical Interrelation.[13]

Color reproduction, made economical through offset lithography, illustrates the current symbiotic relationship between mapmaker and printer. Before the development of color lithography around 1840, maps were largely colored by hand. Multi-color

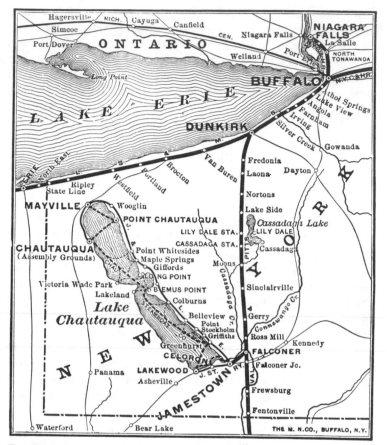

Fig. 6.4.   A map reproduced from an image engraved in wax by the Matthews-Northrup Works, Buffalo, New York.

Source: George H. Benedict, "Map Engraving," *Printing Art* 19 (1912): 205–8 [ref. p. 205].

maps could be produced by laboriously coating different portions of the same printing stone with different inks or by making multiple impressions with different plates in registration, each coated with a different ink. Because lithographic inks tended to obscure any labels or other symbols overprinted with a solid-color shading, line screens were ruled on the surface of the stone to produce various tonal values. Light colors resulted when these parallel lines were comparatively thin or widely spaced, and darker colors resulted when the parallel lines were comparatively thick or closely spaced. With lines spaced 30 to the cm [80 to the in.] or closer, the color was more apparent to the viewer than the pattern of parallel lines. Overprinting two or more patterns with different orientations yielded additional hues, with yellow printed over blue, for example, appearing green. The lithographic transfer process, which obviated the tedious mechanical ruling of line screens directly on the stone or metal plate, was the breakthrough accounting for the much wider use of color after 1875 for maps in geographical journals.[14]

Plastic Film, Scribing, and Feature Separations

In more recent decades, photographic image transfer and stable-base emulsions have greatly simplified the preparation of color maps. A particularly significant advance is *peelcoat,* a stable-base plastic film with an opaque coating that can be cut and then stripped away to open a clear "window" in what is essentially a photographic negative. This so-called *open-window negative* is placed face down over the emulsion of an unexposed sheet of photographic film or paper to form a sandwich, the center of which is another sheet of film (Figure 6.5). On the downward side of this center sheet is a regular pattern of small, closely spaced dots, say, 50 to the cm [130 to the in.]. The three layers of this sandwich are held together tightly by evacuating the air in a *vacuum frame* between a rubber blanket beneath the unexposed film or paper and a sheet of glass above the peelcoat. If the dot screen has clear dots that cover 20 percent of its otherwise opaque surface, 80 percent of the light passing through the open window in the peelcoat will be blocked by the dot screen. The pattern of exposed dots will cover 20 percent of the lower emulsion visible though the peelcoat *mask,* and photographic development and fixing will yield a 20-percent fine-dot graytone.

A closely related development is *negative scribing,* which also employs a thin opaque coating on one side of a sheet of dimensionally stable, clear plastic film. In this case the coating does not

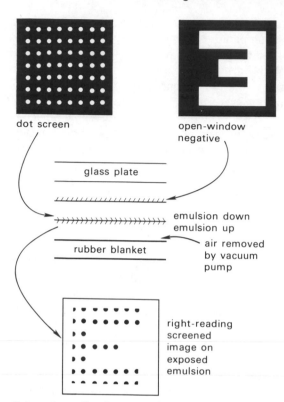

dot screen

open-window
negative

glass plate

emulsion down
emulsion up

rubber blanket

air removed
by vacuum
pump

right-reading
screened
image on
exposed
emulsion

FIG. 6.5. Schematic profile of a vacuum frame (center) showing open-window, peelcoat negative (upper right), dot screen (upper left), and new emulsion, which yields a 20-percent right-reading graytone (lower) for an area defined by a wrong-reading open-window in the peelcoat negative.

peel away. Instead, a drafter draws lines by scratching away the opaque coating with a *graver*. Equipped with either a sharp sapphire stylus, for very thin lines, or a narrow blade, for wider lines, a graver in the hand of a skilled drafter can produce a smooth line of uniform thickness—much more so than with pen and ink. In a sense, the drafter is preparing a negative image that, like a peelcoat, can be used in a vacuum frame to yield a positive image. In this case, though, the open windows are long and thin. Omitting the dot screen produces a solid line; including a screen, as in the sandwich illustrated in Figure 6.5, yields a screened, grayish line.[15] The coating normally is not black but a light orange, through which an image below the scribe sheet can be seen and traced. Yet light with wavelengths that would expose the emulsion is blocked by the *actinically opaque,* orange coating. Special gravers for triple-line and four-line road symbols enhance the technique's versatility.

Scribing evolved from an earlier technique, engraving on a coated glass plate. Graphic artists discovered scribing several times, the first at least as early as 1854. These discoveries were perhaps inevitable, given the early use of glass rather than film as a support for photosensitive emulsions—and the ease with which early negatives were scratched. Many national mapping agencies adopted glass scribing in the early twentieth century. Development of a dimensionally stable plastic film in the United States in the 1930s, its adoption by the U.S. Coast and Geodetic Survey and the U.S. Geological Survey in the 1940s, and its promotion by several manufacturers of graphic arts products are responsible for the current widespread use in commercial and government mapping of scribecoat, peelcoat, and numerous related photo-mechanical products.[16]

A particular attraction of scribecoat is the ease with which maps may be revised. New features need only be scribed onto the original scribecoat. Old features can be deleted by painting them over with an opaque fluid. Errors can also be painted out, and their corrections added on a second sheet of scribecoat. Subsequent exposures in a vacuum frame to both the old and new scribe sheets will yield the intended combined, revised positive image.

Stick-up lettering on *stripping film* is another photo-mechanical advance important in modern mapmaking. Labels for a map can be produced as positive images on sheets of thin, transparent, self-adhesive material from which they may be cut out and affixed in their appropriate positions on a clear plastic sheet. This right-reading *type positive* can be exposed in a vacuum frame to yield a wrong-reading type negative. The image can then be added photographically to those from scribecoats and peelcoats.

Modern storage and revision of map information depends heavily on scribecoats and peelcoats. Features can be scribed on separate sheets of scribecoat and then combined, by their successive contact exposures to the same emulsion. All features of the same variety are often scribed onto a single, unique *feature separation.* Type for place names and other labels can also be segregated by feature. Separation of symbols by feature not only facilitates revision, but also enables the cartographer to employ with a new map selected features drafted for a previous map. Once scribed correctly, a feature need be redrawn only when it or the map scale changes.

Feature separations and dot shading screens may also be used to produce *color separations,* one for each ink to be printed. Figure 6.6 is a schematic diagram describing the development of two color separations, one for yellow and the other for cyan, a shade of blue. Forested areas to be printed green, for instance, may be stored on a single open-window separation negative. This negative can be used in a vacuum frame with appropriate dot screens to place 50-percent and 20-percent images on two separate sheets of film. This first, 50-percent positive image can be used in a further step, to generate for yellow a final negative from which the yellow printing plate can be made. The second, 20-percent positive exposure from the woodland negative is onto the *composite positive* for cyan. Through subsequent contact exposures, this composite positive also receives a 30-percent positive image of the interior of a lake, and a 100-percent, solid image of the lake shore and various streams. Again, a final, composite negative is prepared, from which the cyan press plate is made. Printing the cyan and yellow positive images on the same paper in perfect registration yields green woodlands.

Another advance is *etch-and-strip,* photosensitive peelcoat, which obviates having to trace over the shoreline twice. The lake shoreline in the previous example could have been scribed and the scribecoat used as a mask in the contact exposure of a sheet of peelcoat coated with a special emulsion. Developing etches the surface of the peelcoat and weakens the opaque coating exposed to the lakeshore. After etching, the opaque coating representing the interior of the lake can be grasped carefully at the edge, lifted, and peeled away. The resulting open-window for the lake will be in perfect conformity with the shoreline.

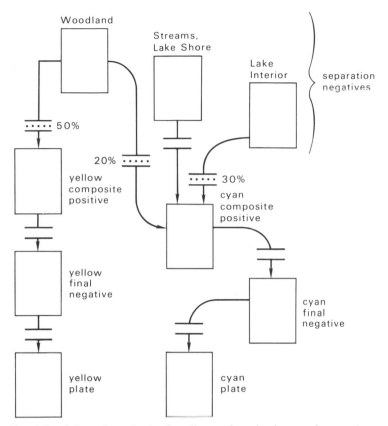

FIG. 6.6. Schematic production flow diagram for a simple two-color map showing color separations for yellow and cyan and feature separations for woodland, lakes, and streams.

The durable polyester film upon which most contemporary cartographic reproduction is based demonstrates the influence upon mapmaking of seemingly unrelated discoveries and technological developments. Polyvinyl chloride, the precursor of modern plastics, is a variant of vinyl chloride, the then-unappreciated result of a 1912 experiment in search of a weatherproof coating for airplane wings.[17] The patent for vinyl chloride was allowed to lapse in 1925, just years before polymer chemists began to develop in the 1930s a variety of synthetic fibers that replaced natural materials in such diverse products as clothing, wall coverings, and radio cabinets.

Every major technological advance has affected mapmaking in some significant way. Patented in 1769, for example, James Watt's steam engine was another prominent impetus to the wider use of maps. Steam provided the power for the more efficient papermaking machines and rotary printing presses of the nineteenth century. The modern equivalent of the Industrial Revolution's steam engine is the digital computer, which can store, retrieve, and display data of many different types, including cartographic. That mapmaking should advance again, because of the digital computer, should surprise no one.

Computer Memories for the Digital Map

A digital map, stored in a computer or on a magnetic disc, might conveniently be described as the electronic instructions for drawing a map. When used to harness and direct the mechanical power of a computer-controlled drafting machine, the informational power of a digital map can yield a traditional graphic, or analog, map. Yet the utility of a digital map lies not only in the generation of visually perceptible images, but also in the ease with which geographic data can be analyzed in the computer and delivered over a wire. Several decades hence, when worldwide and local telecommunications networks conveniently link suppliers and consumers of information, the digital map should be at least as common a cartographic storage medium as the paper map.

The modern digital computer, which provides the impetus for this most recent revolution in cartography, is a technological descendant of the Jacquard loom, an early nineteenth century machine that wove cloth according to a pattern represented by holes punched on a series of paper cards. This form of data storage evolved into the Hollerith card, named after the inventor of a mechanical calculator used to tabulate the results of the United States census of 1880.[18] Each position in which a hole might have been punched is a *bit,* or *bi*nary digi*t*, representing a "yes" or "no" signal. These early punch-card machines demonstrated two tasks important today in computer-assisted cartography: controlling a manufacturing process and carrying out mathematical calculations. Strings of yes-no bits can represent both vector and raster cartographic data, discussed in Chapter 5. Binary data stored for a digital computer employ a number system based upon two, rather than the customary ten. In a string of binary digits, the rightmost digit is the ones place, and the next position to the left is the twos place. Subsequent positions farther to the left represent four, eight, sixteen, and so forth. Thus the binary number 10011, with a "yes" only in the ones, twos, and sixteens places, represents the decimal number 19 (1 + 2 + 16). In a list of vector coordinates, if the first 16 bits of a 32-bit string contain the easting and the second 16 bits contain the northing, the binary number

00011010010111100110111000101011

might represent the location (6,750 m E, 28,203 m N). A raster-format map may also be stored as strings of binary digits, with a specified number of digits assigned to each grid cell.[19]

Through the early 1970s, digital data were commonly stored on punched cards, similar in principle to the hand-punched cards used for the 1880 census, or on paper tape. In the 1950s, electronic data processing consisted largely of sorting and duplicating card decks, tabulating sums, preparing lists, and writing checks and bills. Computers were large, often occupying several hundred square meters of floor space. Circuitry was based upon the vacuum tube, and internal memory was limited. In bulk and speed these early computers were less similar to their modern counterparts than to the Analytical Engine, designed in 1833 by English mathematician Charles Babbage, but built only after his death as a scale-model curiosity.

Magnetic memory was a significant early breakthrough in digital computing. Tiny iron donuts called *cores,* each of which could be magnetized with a positive or negative polarity representing a single binary digit, were first used in 1953, in a test at the Massachusetts Institute of Technology, and were adopted shortly thereafter in commercial computers.[20] First used in 1950 and implemented commercially by UNIVAC in 1951, magnetic tape memory is similar in principle to that used with the home audio recorder/player. Bits can be set by magnetizing tiny areas on a thin mylar tape coated with an iron oxide.[21] With information densities of 5,670 bits per cm [14,400 bits per in.] quite common, magnetic tape still is the most widely used medium for archiving and shipping digital data. It can be read or written many times more rapidly than punch cards or paper tape. Data are stored sequentially, and are most efficiently accessed in the same order as written. Cassette tapes, in a twin-reel cartridge similar to the familiar audiocassette, are also used with microcomputers to store data and programs.

Faster still are magnetic disc and drum memories, with information stored on the iron-oxide-and-epoxy-coated surfaces of platters and cylinders, respectively. Data can be accessed at random, without the inefficient backspacing or reading past unwanted information required when data are retrieved in an unpredictable sequence from magnetic tape.[22] Many small computers store data on small flexible, circular or rectangular discs called *floppy discs*.

Solid-state physicists and computer scientists are continually seeking less expensive, denser memories. Particularly promising during the 1970s were experiments with "magnetic bubble" memory, whereby bits of information can be set through the selective positioning of tiny, movable magnetic bubbles housed in a small chip.[23] But silicon-chip, semiconductor memory seems more efficient. Miniaturization with the semiconductor RAM (Random Access Memory) already has yielded microcomputers with a central processor and considerable memory on a thin wafer less than a square centimeter in size. These chips, which can store 64K (65,536) or more bits of data, are best manufactured in a pure vacuum, free of dust particles, which easily foul their fragile electronic circuitry.[24] An orbiting factory, well above the Earth's comparatively contaminated atmosphere and served by the Space Shuttle, might eventually be economical for the low-cost manufacture of highly reliable semiconductor memories.[25] Given the accelerating growth of digital information and the mounting demand placed on computer memories, scientists are predicting the likely development decades hence of a biological computer with the impressive storage capacity of the DNA molecule and the human brain.[26]

Silicon chip memory has several advantages over magnetic tape and disc memory. Because its elements can be very close together, a microprocessor uses little electric power. In addition to sharing the virtues of compactness, easy shipment by mail or express, low cost, and high reliability, semiconductor memory permits the more rapid retrieval of information. The mechanical inertia in positioning the read head of a disc drive above the appropriate track on the platter can noticeably delay processing if data are to be extracted from widely separated locations on the disc. Because the chip is a random access storage device with no moving parts, data can be retrieved far faster than from magnetic tape and appreciably faster than from magnetic disc. Its portability and ability to store computer programs as well as data has led to the extensive use of the plug-in ROM (Read Only Memory) as a convenient vehicle for distributing software. With both the material substance of hardware and the informational character of software, the ROM might well become a common vehicle for distributing not only digitized maps but also algorithms for their analysis and display.

Semiconductor memory often pales when confronted with the enormous memory requirements of cartographic data processing. Consider, for example, the map displayed on a VDU with 512 by 512 picture elements and up to 64 different colors. A simple raster organization of this picture would require 262,144 (512×512) memory cells of 6 bits (64 is the 6th power of 2) each, for a total of 1,572,864, or 1,536K, bits. An advanced, 64K-bit RAM can accommodate only a fraction of such a map. Even though a compressed scan-line data structure, as described in Chapter 5, might reduce the memory requirement to as few as 300K bits, it is often more efficient to transport and store the data and to recreate the map when needed. Storage of the coverage data alone is particularly efficient for choropleth and other simple statistical maps for which the computer need store but a single copy of the relatively demanding base map.

Not all maps are simple and easily regenerated, though. The efficient cartographic display of land-use and geologic maps, in no way tied to a common set of county or census-tract boundaries, requires a high density, rapid response data store. A particularly promising solution is the *videodisc,* developed for showing mov-

ies, concerts, and other entertainments on the family color television set. The typical disc has about 50,000 tracks, each producing
one complete frame on the screen and thus providing sufficient
memory for one picture on a medium-resolution computer graphics system. Because the surface of the disc is not flexed, as with
video tape, and because the tracking head never touches the disc,
there is no surface wear. Unlike with phonograph records or magnetic tape, frequent retrievals of the same information will not
damage the recording.[27]

Most videodisc systems code the picture as analog information.
First the program is recorded on videotape. These signals are then
used to control the intensity of a laser beam that exposes the
photosensitive surface of a rotating glass disc. The laser light
exposes a series of elongated, gently curving dots, slowly spiral-
ling inward. The length and delicately detailed shapes of these
dots contain as analog code the relative intensities of the three
additive primary colors of the television screen, blue, green, and
red, as well as the volume and pitch of the sound. Position along
the track corresponds to position on the screen. This glass disc is
a master, to be used in a nickel-plating process to produce molds
with which multiple copies are stamped and then given a reflective
coating. The analog signals for the intensities of blue, green, and
red can be decoded by a laser beam and photodetector in the home
viewer's videodisc player and used to recreate the scene and sound
recorded on the original video tape.

A variety of videodisc systems have been developed, mutually
incompatible for the most part. Most detect reflected light, and
after one side is played, the disc must be turned over. Some detect
transmitted light, and are not turned. One approach, with the
signal digitally encoded by a pattern of tiny dots of uniform size,
permits a longer playing time, up to 60 minutes per side, and is
thus particularly promising. Digital videodiscs would seem especially suited to computer graphics systems, which might rapidly
access images at random from various portions of the videodisc.

Its enormous storage capacity hints at the great potential of the
videodisc in mapmaking. The suggestion that all 18 million books
in the Library of Congress might be stored on 100 videodiscs
bodes well for a compact collection of flat maps and atlases,

although map symbols obviously place a greater strain on the VDU than on the sheet of paper.[28] Moreover, map data on videodiscs would be oriented largely toward their direct display, with little opportunity for manipulation by the user. Efficient retrieval would require that the videodisc, or a supplementary machine-readable file, contain a geographic index of place names as well as a detailed cross-reference index describing the location, thematic content, and scale of each frame. Display software might assist the user in exploring the content of the videodisc and in retrieving an appropriate program of map displays and descriptive text. Maps thus might be viewed for a single area at different time periods in sequence, by different themes, or in increasing or decreasing level of detail. A home computer linked to a component television system with a videodisc player could place a sizable map collection in millions of homes.[29]

Mapmakers with sophisticated graphics processors will no doubt develop animated maps to explain the settlement of a region, the evolution of a crisis, or the course of a war. Programs integrated with appropriate commentary and developed for use in the home with unsophisticated videodisc players might enhance the use of maps and other graphics in mass education and even in entertainment. After all, like the book or the film, the map can be fictitious as well as factual. Neither comic books nor encyclopedias will ever be the same.

Graphic Hardware for the Map Publisher

Modern computer technology can contribute significant cost savings to the large publisher of conventional maps. Particularly important is the ease with which computer graphics systems, described in Chapter 5, can make corrections to a geographic data base, generate intricate symbols with precision, and redraft rapidly a fresh, up-to-date copy of a map. VDUs enable the cartographer to inspect corrections and preview new designs. Large flatbed plotters can draw finished maps much more rapidly and accurately than the human drafter—plotting machines, after all, can work a 24-hour day with neither fatigue nor distraction.

Plots can be in ink or on film. In a dark room a light spot can draw on a photosensitive emulsion. Turning the light on and off is equivalent to lowering and raising a pen. Widening or narrowing the aperture yields a thicker or thinner line. The *lighthead* might also have a computer-controlled sliding lettering mask, with negative images of all letters, numbers, and special symbols. This mask can be positioned so that a flash of the plotter's light will transfer the image of a specific character to a designated location on the film emulsion. In this way, the plotter can expose crisp letters with a carefully programmed orientation and spacing (Figure 6.7). With appropriate controlling software, map labels might once again receive the careful placement of the skilled engraver and thus restore the aesthetic quality forced out by the hurriedly placed stick-up type of twentieth-century mapmaking.

Whatever his haste, the human drafter is no match for the modern drafting machine. Along a straight line, the pen, scribing tool, or lighthead might move at a speed of 50 cm [20 in.] per second. Symbols may be positioned to the nearest .01 mm [.0005 in.], with an accuracy of ±.025 mm [.001 in.]—substantially less than

FIG. 6.7.   Examples of maps plotted with a photohead. Example at left is from a 1:5,000-scale urban area map drawn by a Gerber Scientific 4442 high-speed photoplotter for the Institut Geographique National, Paris. Example at right is from a 1:12,000-scale hydrographic chart drawn with a photohead by a Gerber Scientific 4477 flatbed plotter for the Australian Admiralty.

Source: Courtesy Gerber Scientific Instrument Company.

the width of a narrow line. Plotting surfaces of 180 by 300 cm [72 by 120 in.] can accommodate the largest map sheets and folded street maps.[30]

Even more rapid is the raster-mode laser-beam drum plotter, which uses a high-resolution laser beam to plot on a sheet of photographic film mounted on a drum rotating 1,000 times a minute. The laser-beam lighthead migrates down the drum, parallel to the axis, advancing slightly with each rotation to expose a new line of the raster image. The size of the dot flashed by the laser can be varied, as can the width of the scan line, and several adjoining lines can be exposed simultaneously to expedite plotting. With a resolution adjustable to as many as 800 dots per cm [2,000 dots per in.], the laser plotter can produce dot-screen patterns for color separations as well as smooth line symbols, intricate point symbols, and crisp type (Figure 6.8). The laser plotter might take only about 30 minutes to generate a single color separation final negative, from which a press plate can be made. Computer software supplied with most sophisticated laser-beam plotters can readily convert vector images, such as contours and street grids, to raster mode.[31]

When used as a scanner, the drum plotter can also capture digital information from paper maps. A printed or manuscript map can be mounted on the drum. A vacuum pump withdraws through tiny holes any air trapped between drum and map. The scanhead, like the lighthead, moves parallel to the axis of the drum (Figure 6.9). The intensity of the light reflected from the map is measured and recorded on magnetic tape or disc as a digital reading between, for example, 0 and 127. These readings can be stored in only 7 bits ($1 + 2 + 4 + 8 + 16 + 32 + 64 = 127$). The spot for which reflectance is measured might have a diameter as small as .025 mm [.001 in.] or as large as .25 mm [.01 in.]. Coordinating the width of the scan track with the frequency of the reading can yield a grid of square cells. If the scanned map were printed on white paper with black ink, a *slice level* of, for example, 50 might differentiate cells with ink on the source image from those without. The inkless, non-image cells will have reflectances well above 50, and cells in the center of a dark line will have reflectances near zero.[32]

Scanning yields raster data. Codes identifying map features

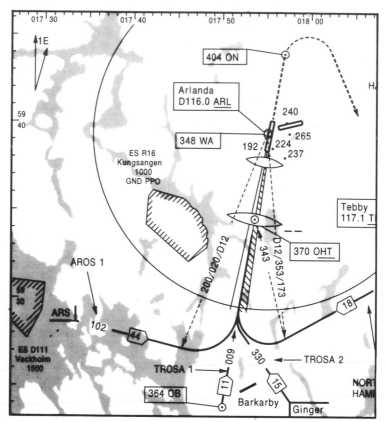

FIG. 6.8. Example of a map plotted with the Scitex Response-250 laser-beam drum plotter.

Source: Courtesy Scitex America Corp.

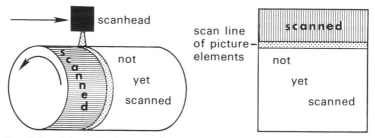

FIG. 6.9. Schematic diagram of a drum scanner (left) and the scanned raster image (right).

are assigned to lines or areas by an operator at an interactive graphics terminal. If the data base of the processing algorithm requires vector data, line-following software can track linear features through the grid and generate lists of coordinates.[33]

Vector data can also be captured manually, by following lines on a digitizer with a hand-held cursor. Point coordinates can be recorded by pushing a foot switch when the cross hairs on the cursor intersect above the intended point. Digitizers can also record points at a constant time or distance interval as the operator tracks along the feature. Codes identifying the features can be entered through a keyboard or a keypad on the cursor, a menu on the digitizer similar to that described in Chapter Five, or a voice decoder programmed by the operator to recognized spoken codes and commands.[34]

Optical scan digitization is preferred to manual digitization. A scanner is faster and more accurate—and the map publisher who cannot afford to own one should consider paying another firm with a scanner to do the work or lease the use of its hardware. With a 50 by 40 cm [20 by 16 in.] map, scanning might require one hour, code assignment 12 hours, and raster-to-vector conversion one hour. For a moderately complex map, for example, a typical topographic map, a human operator might identify and track all features in 250 hours and edit the file in another 150 hours. Because of involuntary nervous twitches and other quirks, predictable and unpredictable, the manually digitized vector file will more certainly contain numerous "spikes" and other "glitches." Scan digitization with software conversion to vector mode, where needed, is more accurate and, for publishers large enough to fully use the equipment, more economical.[35]

Software to convert from vector format to raster mode is every bit as important as software for raster-to-vector conversion.[36] Although some features are more readily symbolized by such vector symbols as contour lines and double-line roads, maps on which they are shown can be generated more efficiently on raster-mode display devices. High-resolution laser-beam drum plotters, as noted earlier, can produce final color separation negatives. With slight design modifications, laser plotters might also produce press plates directly. The time required to plot a raster-mode map depends for the most part on size and resolution. Content and graphic complexity, highly important in pen-plotter mapping, have little effect on the time needed for a raster plot. A precision flatbed plotter might take six hours to draw a map that a laser plotter could produce in one hour—with four different separations for color lithography!

For small runs, up to several hundred copies, computer-controlled printers can be more efficient than offset lithography for reproducing raster map images. Among the more promising of these electronic printers are the electrostatic dot-matrix printer and the ink-jet plotter. Both processes treat an image as a uniform grid, or matrix, of dots. These dots may or may not be plotted, and a single bit can represent each dot cell. The dots must overlap slightly where solid black is to be printed. Image quality depends upon the density of dots. Resolutions of 100 dots per cm [254 per in.] have been attained, and even somewhat coarser displays can provide clear labels, smooth lines, and detailed point symbols (Figure 6.10). Printing speed tends to be inversely proportional to image quality, although dot-matrix printers are appreciably faster than raster drum plotters and pen plotters. Electrostatic graphics printers available in 1982 could plot as many as 1,360 lines of dots per minute—17 cm [6.8 in.] per minute on rolled paper for a map with a resolution of 80 dots per cm [200 dots per in.].[37]

Some electrostatic printers feed paper from a roll past a row of closely spaced metal styli, the spacing of which determines the resolution of the printed image. A grounded platen on the opposite side of the paper permits a flow of current from each stylus through the paper. This current places a small charge on a dot between stylus and paper. The paper then passes through a tray of dark or colored powder, or *toner,* grains of which are attracted to the charged parts of the paper. Subsequent heating by a hot lamp fuses the toner to the paper and fixes the image.[38] Some electrostatic printers produce plots in several colors by using a number of rows of styli, each with its own tray of unique, colored toner. Heat-sensitive paper that changes color chemically, without toner, may also be used.

Ink-jet printing is based upon a 19th century observation of the English physicist Baron John W. S. Rayleigh: if vibrated rapidly,

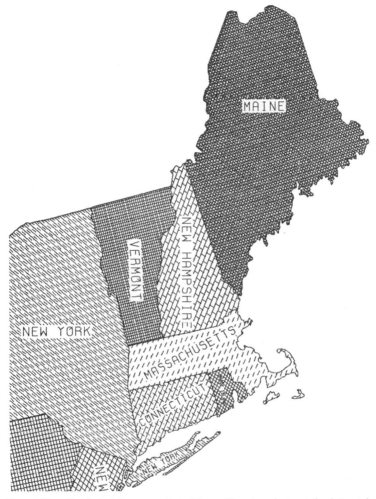

FIG. 6.10. Example of a map plotted by a Versatec electrostatic dot-matrix printer/plotter. Portion shown is from a larger map of the United States plotted with a resolution of 80 dots/cm [200 dots/in.] on paper 56 cm [22 in.] wide.

Source: Courtesy Versatec, a Xerox company.

at about 117,000 cycles per second, a jet of ink forced through a small nozzle breaks apart into tiny, uniformly spaced droplets of consistent size. An electrode placed near the tip of the ink jet transfers to each droplet an electric charge, the strength of which can be varied (Figure 6.11). Pushed by its momentum at the nozzle into an electrostatic field between two deflector plates, the charged ink droplet can be deflected anywhere along a narrow band that forms a line of the printed image. The amount of deflection depends upon the charge, determined by the voltage of the electrode at the time the droplet was formed. Varying this voltage can position ink droplets anywhere along a row of the image or deflect them into a gutter. The nozzle produces a continual stream of dots, and the gutter collects for later use the ink of unwanted dots. After all dots required for the row are in place, the paper is advanced to the next row. Multiple jets can be used for longer lines as well as for different colors.[39]

The variety of electronically controlled print-out devices seems limitless. Several manufacturers have introduced photographic laser printers designed to serve business offices not only as copiers but also as hard-copy units for word processors, computers, electronic mail, and long-distance facsimile transmission of text and diagrams. Speeds exceed 43 correspondence-quality letter-size pages

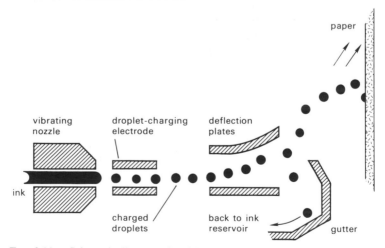

Fig. 6.11.   Schematic diagram of an ink-jet printer.

per minute, with a resolution of 120 dots per cm [300 dots per in.].[40] Designs as simple in concept as adding blue and yellow to the two-tone, black-and red typing ribbon might eventually provide even owners of personal computers with inexpensive hardcopy color maps.[41]

The proliferation and continual revision of digital map data suggests microfilm as an inexpensive medium for preserving snapshots of maps at selected stages in their evolution. In addition to its archival role, the microform edition could be a significant secondary publication if many users had access to microform readers. Microform distribution is promoted by the computer-controlled COM (Computer-Output-Microfilm) recorder, which transfers digital information to film by photographing the screen of a small CRT or by writing directly onto the film with a fine beam of light controlled by a microprocessor.

Although most COM recorders are intended only for alphanumeric data, specially designed COM units can plot maps and diagrams with high resolution.[42] Enlargement twenty or more times does not appreciably degrade the image. In fact, early successful attempts at computer-assisted production of lithographed color maps employed color separations enlarged photographically from images plotted on a COM recorder.[43] Hard-copy graphics units that plot on standard color transparency film will be particularly useful in preparing microform maps as well as color slides to illustrate lectures and briefings.

As with the non-electronic printing of the nineteenth century, the electronic printing of the late twentieth century will evolve through much experimentation. Despite considerable trial and error, inevitable progress is now indicated on two fronts—traditional, centralized mass duplication in the printing plant and "one-off," demand printing in the home, office, or local news store. Accelerating change will demand from the mapmaker more skill and vigilance than ever in selecting and adapting publishing technology.

Glass Threads, Networks, and Videotex

Advances first in transportation but more recently in communications have greatly increased the speed at which a map can be carried from publisher to consumer. In the Middle Ages, a messenger on horseback could cover perhaps 100 km [62 mi] in a day if the weather were good. A sailing ship might cover 200 km [120 mi] in the same 24 hours—or make little progress at all if becalmed or diverted by a storm. In the nineteenth century, with the advent of the railroad, a map, like most other merchandise, could be shipped a thousand miles in a day or two. Modern jet aircraft now provide inexpensive overnight delivery over still greater distances. Yet a network of communications satellites and ground lines could transmit the same information halfway around the world in seconds.

Historians of cartography have given little attention to mass transportation and communication.[44] Vehicles and ships, it would seem, have had little apparent effect on the look of maps. A totally different attitude, though, is likely among future historians of cartography who will be confronted by the profound effect on maps of electronic communications. Maps sent by wire or electromagnetic waves, after all, need not be printed. Moreover, the increased utility of up-to-date maps forwarded by electronic mail will most likely lead to the demise of many publishers of visually attractive but less informative, traditional maps. The map user should always be more attracted to an aesthetic but functional design. However, by changing the standards for what is considered functional and informative in maps, modern communications are changing the constraints and objectives—not to mention the media—with which the cartographic designer must work.

Electronic publishing raises the question, Where will the map be designed, and by whom? If the transmission rate is too slow to permit sending completed pictures, the map data alone might be sent to the customer, who then will compose the map display on his own graphics system. In this case, design principles might be embedded in the home user's mapping software. If so, the master cartographer of the future will exert a much greater professional influence through canned programs than through original artwork. Yet, if the transmission rate is high, the user can readily receive maps with all graphic details selected and balanced by a skilled designer. Despite the likelihood of individually tailored, one-of-a-kind maps—a "de-massified" cartography, to borrow a term

from futurist Alvin Toffler—high transmission rates will still promote works of cartographic art, forwarded as graphic messages
rather than geographic data.[45]

Transmission rates and charges will affect both the variety and
timeliness of map data and authored maps available through an
electronic information network. After all, the efficiency of a telecommunications link depends upon not only how fast but also
how much data can be moved. Although electronic signals move
through air or wire at essentially the speed of light, the various
electronic media differ greatly in the speed with which a particular
message can be carried without distortion. To remain intelligible,
the various parts of a message must be separated from one another
in time or transmitted over several channels. A television signal
can be more complex, and its transmission rate higher, than a
telephone signal, which is limited by the electrical resistance and
capacitance of metal wire. The television signal, which has a
greater range of frequencies, is said to have a greater *bandwidth*.

With a digitally encoded signal, the capacity of a link can be
measured as the number of bits carried per second. The comparable bit rates of a variety of communication activities permit
some interesting comparisons. A good typist, for example, enters
information through a keyboard at about 40 bits/sec, a normal
person reads text at 400 bits/sec, a voice-grade telephone channel
can transmit 56,000 bits/sec, and a color television channel can
carry 92 million bits/sec.[46] Wire cable limits the signal rate, and
a pair of well-shielded, low-resistance wires cannot carry more
than 6 million bits/sec. Surrounding a central single-wire conductor with an outer, coaxial, cylindrical conductor, and separating
the two by air and plastic insulators, can increase the transmission
rate. Coaxial cables, used by cable television networks to carry
simultaneously 40 or more channels, can transmit much higher
frequency signals than a wire pair, but with increased electrical
resistance. Repeater stations are needed to amplify the fading
signal.

Cables of thin glass fibers, rather than metal wires, are particularly promising for the cost-effective transmission of digital picture data. Fibers of high quality optical glass with a diameter of
about 0.1 mm [.004 in.] and a lower refractive index near the

surface can transmit over great distances laser signals pulsing, in theory at least, at the frequency of light. Like metal wires, these glass fibers are flexible and can turn corners. Energy loss is low, and repeaters can be more than 10 km[4 mi] apart. A single fiber might transmit one billion bits/sec, but rates of 100 million bits/ sec are more practicable. Thousands of fibers can be bound to- gether yet still occupy a diameter no greater than coaxial cables now in use. Data transmission technology, already far ahead of the computer's ability to generate and receive information, should encourage the development of improved image communication, including high-resolution, wall-size television screens.[47]

Land-based telecommunications links will probably exist prin- cipally for the local distribution of signals, with a small number of stationary, high-capacity satellites handling most long-distance communications. Attitudes toward satellites have changed greatly since the first commercial satellite was launched in 1965. From its early perception as a medium for reaching inaccessible places, the satellite has evolved into a facility able to carry a variety of signals for many users: inter-continental and long-distance tele- phone calls, television programming, digital data, and electronic mail. The Space Shuttle will lower the cost of launching heavier, more powerful satellites, with large, reliable solar panels. Micro- circuitry provides for a higher traffic volume and the enciphering needed for confidential business communications. *Transponders,* which receive and retransmit an amplified signal on a different frequency, increase further the power of the satellite by permitting the use of smaller, less powerful and less costly earth stations. Indeed, much of the present terrestrial network might be rendered obsolete by individual rooftop dish antennas, and corporations with extensive investment in terrestrial long-distance links might attempt to retard the full development of efficient satellite-oriented telecommunications.[48]

Whatever network does evolve, *packet switching* will be impor- tant in its efficient operation, particularly for data that can be sent in bursts, as from a keyboard terminal. A *packet* is a group of bits with a control header identifying the type and length of message and the sender, a destination address, the data message, and sev- eral *check bits,* used by the receiving terminal to check for inaccurate transmission. Compared to cartographic requirements, individual packets are small; for example, hundreds of packets carrying only 2,048 bits of data might be needed to transfer a single, moderate-resolution CRT image. Yet transfer can be fast and efficient between distant terminals that communicate infrequently. The message is assembled at a local or regional packet-switching exchange, perhaps carried there slowly, character by character, over telephone wires. The exchange divides the message into packets, adds the appropriate headers, and forwards the packets through the network to their destination (Figure 6.12). Each node receiving a message examines the address and retransmits the packet if the node itself is not the exchange servicing the specified destination. Depending upon traffic at the time of use, transmission will generally be noticeably slower than with *channel switching,* whereby a direct channel is cleared between two points. On the U.S. Department of Defense ARPANET (Defense Advanced Research Projects Agency Network), packets are delayed an average of only 0.2 sec. Packet switching could be useful for responding to the requests of users for specific data.[49]

Important factors in designing networks for cartographic transmission include the volume of data, delay time, differing directional requirements for two-way communications, and the relative numbers of senders and receivers. Videotex systems recently developed for providing screens, or *pages,* of information to home television viewers illustrate these considerations. Two types of videotex system are in use, teletext and viewdata.[50] *Teletext* systems, such as the British Broadcasting Corporation's Ceefax, is a one-way system broadcasting information with the television picture signal. Each channel carries a hundred pages of information, transmitted four a second in the unused portion of the picture signal, the vertical blanking interval. The user selects a page by punching a number into a keypad attached to her television set, which then grabs the particular page and displays it on the screen. Index pages point to other pages with related or more detailed data. The user need wait no longer than 25 seconds for the requested page. Adding more pages to a standard picture channel would only increase the average delay, but special channels dedicated to teletext and either carried by cable or broadcast could

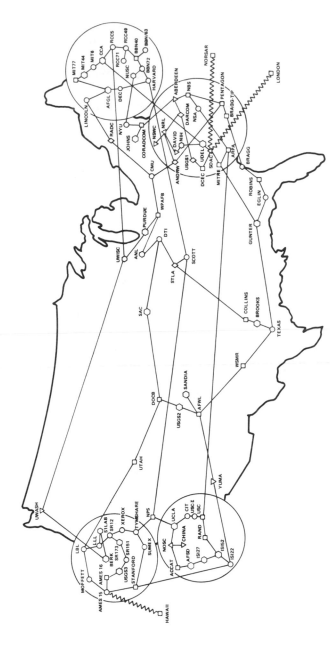

Topology of the ARPANET, January 1982

FIG. 6.12. Schematic map of the ARPANET showing links between nodes. Zig-zag links represent satellite connections.

Source: Courtesy Defense Advanced Research Projects Agency.

support many more pages. Teletext systems might compete with or supplement the daily newspaper and even carry advertising.[51]

Viewdata systems, which support two-way communication, are more versatile than teletext. Competing but technically incompatible systems have been developed in England, France, Canada, and the United States. Among the more promising is Telidon, developed jointly by private industry and the Canadian Department of Communications. A data base of 100,000 pages is planned. Information may be requested and received over telephone lines, coaxial cable, or a fiber-optics network. As with teletext, requests are entered through a hand-held keypad and displayed on a color television set. Because coaxial cable can transfer high-resolution pictures more efficiently than voice-grade telephone lines, viewdata systems are particularly suited to two-way TV cable or to separate one-way transmissions of requests through the telephone network and pictures through TV cable systems.[52]

Graphic resolution is a particularly important concern for cartographic publishers. Map symbols cannot be accommodated as easily as alphanumeric characters. Most videotex systems are designed solely for the dissemination of text, with each line of the transmitted "picture" expanded to fill several lines of a relatively coarse picture on the viewer's screen. For a while, at least, the success of a system will depend on its ability to provide subscribers with a wide variety of pages and to transmit rapidly the requested data. These capabilities conflict, unfortunately, with high-resolution graphics. Technical standards adopted by agencies regulating electronic communications will greatly affect videotex's future as a cartographic medium.[53]

Telidon, an acknowledged leader in videotex graphics, makes a fuller use of existing telecommunications links by transmitting maps and other graphics in a packed code.[54] A page is stored and transmitted as a series of picture description instructions (PDIs), which treat the image as a set of standard geometric elements such as lines, circles, and arcs.[55] A microprocessor in the terminal then constructs a video display by generating these picture components in their specified positions. By specifying the sequence in which map elements are added to the screen, the cartographer can control the user's cognitive processing of the map display more fully than with the traditional, printed map.

Information networks such as viewdata and related services provided to households through improved telecommunications will influence substantially the abundance, availability, timeliness, and content of maps. Like the electronic calculator, cable television, and the videogame, electronic publishing will become widely available in the near future because of lowered prices and a wider variety of useful services. James Martin, a technical journalist prominent in computers and telecommunications, lists 107 different services that might be offered through existing telephone lines and television cable systems.[56] Martin's major headings are passive entertainment, people-to-people communication, interactive television, still-picture interaction, monitoring, telephone voice answerback, home printer, and computer terminals (including the viewdata television set). Listed under "home printer" is "obtaining travel advice/maps." A wider role for maps is suggested by such other services as city information, boating/fishing information, weather forecasts, real estate sales, and job searching. Yet most of the services Martin lists can be satisfied by text alone. Although maps clearly have a role in the telecommunications-information revolution, as in the development of printing and photography, this role is clearly secondary and more passive than active. While not a prime innovator, the mapmaker must be an efficient and innovative adaptor.

Maps as Software

The Electronic Transition seems destined to progress to a stage at which maps are seldom composed on paper or similar graphic media. At present much effort is devoted to data capture, the conversion of geographic information to digital form by scanning or digitizing printed or manuscript maps. With time, almost all cartographic data will be captured electronically, at their source, through photogrammetry or satellite remote sensing, or from computer systems for processing census results and other administrative data. Digitizers and scanners will be much less common than at present, and emphasis will be on the revision, dissemination, and use of the information base. A related transition toward the electronic map is the evolution in display hardware from the digital-mechanical printer and map plotter of the 1950s and 1960s through the digital-photographic film-writer and laser plotter of the later 1960s and 1970s to the instantaneously transmitted, electronically plotted interactive displays and videotex systems of the 1980s.

A prime casualty of cartography's Electronic Transition will be the attitude that the map is a printed product. As with most other forms of published information, and perhaps even more so than for books and magazines, maps should be viewed as software rather than material objects.[57] Moreover, there are at least three types of map software: map data, mapping algorithms, and finished maps. In many cases the latter can be generated from the first two, particularly if the data are much more transportable and informative than the finished map. A map publisher might deal in any or all of these types of map software.

While it might be too soon to forecast the demise of the paper map, paper clearly will become less important as a storage medium and vehicle for geographic information. To be sure, forecasts are error-prone, with or without a crystal ball. Innovations can appear far more promising to the seer than to the user, and foresight might well be little more than tunnel vision. The 1894 writer who predicted the replacement of the book by the phonograph and Edison's kinetoscope, for instance, ignored an average reading speed well above the average speaking rate.[58] But telecommunications and computer information systems clearly have the support of business and government, and seem able to overcome whatever grassroots resistance there might be to computers and electronic innovations. The economic battle seems to have been won, and the political fight is progressing nicely, especially in the schools. Computer instruction in elementary and high schools will with time surely diminish most resistance to electronic publishing.[59] The paper map might not be dead, but the digital map will greatly alter where, how, and when paper maps are printed.

Early training with computers and electronic information systems should equip the next generation of mapmakers and cartographers for an unprecedented amount of experimentation with hardware, data structures, and algorithms. If the history of cartography has anything relevant to say about the future, the computer will duplicate some of the impacts of printing noted by Robinson.[60] Although the intermediate periods might be shorter and less clearly defined, successive stages of media trials, essential separation of mapmaker and computer specialist, automation and innovation, and technical interrelation should all occur. At present, it would seem, we are in a period of automation and innovation. Media trials predominated during the 1950s and 1960s. Despite some personal contacts during the 1970s at professional meetings and in universities, cartographers and computer graphics specialists in computer science have separate technical journals and are still largely segregated. The current compartmentalization of academic disciplines and professional associations suggests that the final period of technical interrelation noted in Robinson's study of map printing has yet to arrive. But whether cartography and computer graphics are "interrelated" or not, the current and increasing dependence of cartography on electrical and computer engineering should be abundantly clear.

CHAPTER 7

Summary and Conclusions

Cartography's Electronic Transition presents new problems at the same time that it introduces new possibilities. To assume that maps will be better—more accurate, more timely, more accessible, more aesthetic, more tailored to user needs—simply as a result of high technology is unreasonably naive. New and evolving technologies for handling map information must be adapted carefully and selectively. The greatest challenge is that of management and organization. Geographic information systems are complex; unless the parts fit together precisely, a system will be ineffective and perhaps more cumbersome than a small army of clerks retrieving paper maps from thousands of large, flat filing drawers. The potential for better maps has never been greater, yet neither has the threat of dismal, expensive, embarrassing failures.

Nor can modern mapping technology obviate the age-old subjectivity and selectivity that rests, as always, on sound judgment and conscientious ethics. Maps have a tremendous capacity for deceptive posturing: they look authoritative and are not readily challenged by lesser authorities, however suspicious. Consider the frustration of the Commissioners for the Internal Improvement of the State in their 1814 report to the New York legislature:

> . . . Disdaining to consider the actual state of things; whenever map-makers trace a stream, they find a military and commercial highway. Should there be a want of water, it is supplied by their depth of intellect; should the surface be covered with ice, it is thawed by

their warmth of imagination. To contend with such men is no easy task, for they make facts as they go along, and reasons they disclaim; insisting, that whatsoever they think proper to approve of is sublime; whatsoever they think proper to dislike, is absurd. From these decrees, pronounced with an air of censorial gravity and the contemptuous smile of superior intelligence, they admit of no appeal.[1]

Computers, satellites, and telecommunications have not erased the potential for careless arrogance or even well-meaning ignorance. Instead, the problem is compounded by our high technology, which appears to be pushing ahead far faster than the society, institutions, and management techniques that must adapt and control it.

Summary of the Chapters

The preceding chapters identify a wide variety of areas in which modern technology is improving the effectiveness of maps. Navigation, discussed in Chapter Two, is one facet of map use with a long history of generally effective management, perhaps because mistakes could lead to lost lives, property damage, military defeat, or national disgrace. It is also an area experiencing considerable technological "trickle down" from the development by the military of geodetic positioning and missile guidance systems. A constellation of navigation satellites might eventually eliminate most of the field work required in control surveying as well as provide guidance for nuclear warheads. The extent to which navigation technology is made available for peaceful, civilian uses will depend on government policy and the ability of management to control crucial, classified design and access specifications. But some transfer is inevitable if the ideas and their implementation are obvious, practicable, and profitable. Already the principles of inertial navigation have been applied in an automobile-mounted automated map display. A dot of light on a CRT screen charts the vehicle's path along the road network shown on one of several available transparent maps that can be positioned over the screen.[2] This innovation, heralded years earlier as one of the special effects in a cinematic thriller, can become as commonplace as the CB radio or the toaster oven if the public finds it useful.

Surveying, treated in Chapter Three, is another area that has benefitted from the military's need for accurate geographic and geodetic information. It also underscores mapping's position as a publicly owned and operated utility, similar in some ways to America's highways and Europe's railways. Mapping is indeed an important part of any nation's infrastructure, and geodetic control and large-scale base mapping need the uniform standards and coordination of a central government.

Photomapping and remote sensing, the theme of Chapter Four, is yet another area in which military and civilian objectives mesh successfully. Although some equipment and sensor data are classified—unfortunately, not always for the most valid reasons—the wide range of civilian applications and their enthusiastic acceptance by earth scientists and resource managers is likely to assure the continued availability of remote sensing products to businesses, local officials, and scholars. Nonetheless, in the United States the future role of government is uncertain: satellite remote sensing, as indeed photogrammetry since almost its inception, can be operated at least jointly with the private sector. Over a decade of cooperation between NASA and several private operators of communications satellites demonstrates the efficacy of collaboration between sectors, as does the extensive use of nonfederal, contractor personnel in the operation of the EROS Data Center. If the control survey network is to be funded as infrastructure similar to the highway system, then satellite remote sensing will probably be owned and managed very much like public utilities for hydroelectric power generation and telephone service. A monopoly might not exist, but the right of public access at a reasonable cost must receive as much protection as investor profit.

Remote sensing data and other information collected by government agencies and private firms is valuable in proportion to its organization for effective retrieval and analysis. Decision support systems, the focus of Chapter Five, are benefitting from a multitude of innovations in the computer industry, especially in data entry, storage, and display. Many hardware manufacturers, large and small, are involved, and carefully conceived, widely accepted standards are needed to assure component compatibility. Equally if not more important is the need for efficient, flexible, well-documented, and theoretically sound program software to provide a reliable link between data software and display hardware. Complex computer programs linking a variety of spatial data bases and making operational a wide variety of graphic and analytical algorithms are a significant challenge to managers and researchers in both the private sector, where much of the work will be done, and the public sector, which controls much of the data.[3]

Map publishing, so long linked with map printing, is discussed in Chapter Six, which points out that centralized printing establishments will no longer play as vital a role as before in the distribution of map information. Telecommunications networks, videodiscs, videotex systems, and home computers are but a few of the new information media that will revolutionize map publishing. A wider variety of maps should be available, these maps should be more timely than at present, and their users will no longer need to settle for a single map, rigidly freezing on paper the whims of its designer. Yet a greater variety of maps will probably be printed, even as printed maps decline as a proportion of all maps viewed or searched.

Publishing is far more than printing, warehousing, and shipping. The need remains for someone to collect, review, and organize map data, to make others aware of the data's existence, and to distribute these data to interested buyers. Through modern telecommunications the potential exists for a small but competent self-publishing geographer-designer to advertise widely a useful product, compete with the better endowed subsidiaries of multinational corporations, and earn a fair return on labor and intellect.[4]

Yet a rosy future is by no means guaranteed. Digital maps might fare no better with the information system manager than have their paper counterparts with the head librarian. As in the technological tempest mapping has recently entered, mapping will be a minor factor in the tele-publishing network upon which it will depend. Map users must lobby to protect their particular interests, while adapting to the dominant demands of nonspatial information.

Problems for Public Policy

Mapping has both a unique and a shared interest in public policy. In matters of privacy it shares some of the concerns raised about videotex and other electronic media as possible instruments of excessive government control. Linking together various computer records systems may shift the balance of power between a government and its citizens.[5] Suppression and slanting of information released to the public by politicians is a particular concern, given the authoritative power of the map and the easily embarrassed psyches of some public officials. High-resolution remote sensing systems, for example, can collect substantial amounts of potentially sensitive environmental data that polluters would prefer be kept from an alert and apprehensive public. Social indicators compiled by the Census Bureau, economic research units, and health and welfare agencies are rife with value judgments, particularly those of technocrats who prefer to deal with numbers rather than with people and the quality of life.[6]

Copyright is another widely shared policy concern. Perhaps because of their highly stylized artwork and the often questionable originality of their information, copyright has not provided the paper map the same protection afforded literary and dramatic works.[7] As digital data, though, maps have much in common with computer software, proprietary information, and electronic games. Digital maps thus benefit from the protection provided by the Computer Software Copyright Act of 1980, which defines a copy as any "material object . . . in which a work is fixed by any method now known or later developed, and from which the work can be perceived, reproduced, or otherwise communicated, either directly or with the aid of a machine or device." ROMs and other magnetic representations can be copyrighted, and appreciable uniqueness or originality is not required.[8] Yet this protection extends only to outright copying, insofar as only patent protection or a legal agreement between user and supplier can protect a program or algorithm's creative originality.[9] Hence map data are more easily protected than mapping software.

Courts and legislatures will, no doubt, have to address new legal issues in both the protection of intellectual property and the liability for damages resulting from misinformation. Protection only against unauthorized copying is afforded derivative maps, such as those produced by a computer program analyzing several copyrighted data bases and others in the public domain. How much protection, if any, should be granted someone whose information is used but not copied? Can a momentary examination by computer be a legally sustainable form of copying even though another identical copy does not result? Regenerating this derivative map from the same sources presumably would not infringe an existing copyright, but this limitation might lead to overly vague explanations, suppression of data documentation, and even deliberately misleading descriptions. Buyers kept partially ignorant of the data's origin might also be asked to sign waivers of liability should their analyses or interpretations produce financial losses. These possibilities have, of course, always been present; digital cartography merely increases the likelihood of damages and litigation.

Related to questions about copyright are issues of the right of access to publicly owned information and the fairness of chargebacks for this access. Seldom charging the full cost of printing and distribution, the United States has heretofore provided map information as a free public benefit to its citizens, industries, and local governments. Great Britain, in contrast, has traditionally held a Crown Copyright for Ordnance Survey and other publicly produced maps, and had charged royalties even for permission to compile derivative maps such as urban street guides. The comparative ease with which transmissions of digital data can be metered, together with the escalating costs of collecting and managing cartographic information, raises the specter—or hope?—of chargebacks and, possibly, full cost recovery.

But what schedule of rates would be fair to both the incidental user such as the sportsman or map enthusiast and the large, frequent, profit-making user such as the public utility or the regional environmental and civil engineering firm? Can the interests of the scholar and the amateur historical geographer be recognized and protected? And can national mapping programs be shielded from blind allegiance to a cartographic Nielsen rating that might lead to topographic equivalents of the soap opera or game show? Coverage of less popular areas and map themes must be provided, as must low-cost map information for the nonprofit hobbyist and the landscape enthusiast. Might not a National Endowment for Cartography be established?

Remote sensing data, as noted earlier, might well pass largely from NASA and the public domain into the hands of a privately owned and operated consortium similar to COMSAT, which operates communications satellites used for long-distance telephone and television transmissions. Another possibility, begun on a trial basis in 1983, is the operation of the Landsat program by the National Oceanographic and Atmospheric Administration (NOAA), which also operates the nation's system of weather satellites. Yet NOAA may be no more effective than NASA as an operational agency, and the United States might well follow Europe and Japan in establishing commercial remote sensing satellites. Even so, some government commitment to purchase a minimum amount of information would probably be needed, as would an arrangement with NASA to place the satellites in orbit.[10] Continued federal support might also be provided for resource inventories in Third World nations, some of which have found Landsat data a useful tool for economic development as well as a basis for closer ties with the United States. And a public operational environmental satellite firm might be restricted from supplying foreign governments or corporations with information of military or strategic economic significance.

The coordination of data exchange is even more important a public policy issue in transfers between federal and state agencies. Difficulties frequently arise from inaccurate or inappropriate data formats, users not aware of or not able to obtain the needed data, federal agencies responding too slowly to requests, and federal agencies refusing to release data deemed confidential or "nonpublishable."[11] Coordination must clearly be organized at the federal level, with a mandated participation by states and localities. Much effort can be saved if delays are reduced, knowledge shared, needless redundancy eliminated, common standards agreed upon and honored, and compatibility assured. Effective federal-state coordination can also promote the highly useful but even more difficult intrastate coordination.

Without coordination at the federal level, federal-state coordination is problematic at best. Despite periodic suggestions that a single federal mapping agency be formed—the idea seems to be discovered about once every decade—no action has been taken and federal cartographic activities are fragmented among 38 separate agencies.[12] Digital cartography and the need for increased integration might provide the ultimate impetus. Indeed, the De-

fense Mapping Agency, formed in the early 1970s from separate
mapping branches in the Army, the Navy, and the Air Force, is a
model of successful consolidation. Combining civilian mapping
activities has the enthusiastic support of the esteemed National
Research Council. An NRC panel recently affirmed the need to
develop extensive, fully integrated digital data bases for geodetic,
cartographic, and marine data.[13]

Yet the retarding effect of bureaucratic inertia as well as the
traditional fragmentation of a civilian mapping establishment that
has functioned for years without political embarrassment suggest
that, like technological progress, change is more likely to come
from outside and above, not from within. What most likely will
be required is a major, cabinet-level realignment of government
operations, with mapping activities consolidated under a Department of Natural Resources. Even then, mapping of economic and
population statistics most likely will remain separate, to be gov-
erned by the nation's statistical policy, not its mapping policy. An
effective and institutionally sanctioned liaison will be needed to
compensate for this perhaps unavoidable obstacle to the full consolidation of federal civilian cartography.

Mapping policy analysts must also be prepared to deal with the
emerging role of private enterprise, particularly the multinational
corporation. Federal, state, and local governments are constrained
by the fixed boundaries of an agrarian culture, whereas private
enterprise can integrate its activities nationally and internationally
without regard to political boundaries.[14] Enterprising private-sector
mapping firms might well arise to organize and manage selected
mapping services for territorial units poorly served by artificial
municipal, county, state or international boundaries. To be sure,
exploitation of marine resources might best be organized not by
governments but by multinational conglomerates. The oceans and
their depths could provide cartography's last unexplored frontier.[15] Further, the possibility of intense private-sector action might
be the catalyst for forming a single civilian mapping agency.

Security and Digital Data

Security is another increasingly important concern confronting
cartographic management: without adequate safeguards, a digital

data base is more vulnerable than the traditional paper map and its reproduction separates. There are two main threats—nuclear attack and the malevolent prankster. The armed terrorist might also be recognized as an intermediate case. The concentration of magnetically stored digital data in a single location without the backup of duplicate files elsewhere, in several secure sites, would be almost an open invitation to attack or tampering. Armed guards, identification badges, closed circuit television surveillance, and personnel security clearances can be useful, to be sure, but limiting security measures to these traditional responses provides no more protection than did the Maginot Line. A single nuclear blast well above the land surface might hopelessly disable the flow of map information and users' abilities to process it, and bored, precocious and amoral preadolescents can penetrate a computer system by terminal and telephone without setting foot on the property. Security is a major undertaking for both local managements and national defense.

For over a decade electromagnetic pulse (EMP) has been recognized as a prime military threat to semiconductor circuitry and telecommunications. A nuclear explosion at 500 km [800 mi] could cover all the conterminous United States with a high voltage wave that would shut down the electric power grid and disable unprotected telecommunications equipment.[16] Because metal wires would conduct the EMP to fragile integrated circuits, fiber optics communications channels would be useful in *hardening* our communications and information storage systems. Hardening might also involve independent power supplies, vacuum tube circuitry, and satellite communications. Even so, satellites are vulnerable to airborne or orbiting laser particle-beam weapons and the explosive-carrying antisatellite satellites, which can be maneuvered nearby and then detonated.[17] Perhaps the best protection for our cartographic legacy, if one can accept the possibility of surviving a nuclear war, is the widespread archiving of duplicate data files deep below the surface and an even more widespread stockpiling above the surface of paper or microform copies. The burial strategy presumably has already been followed, possibly in conjunction with the Federal Reserve Bank's storage of duplicate records in the Carey Salt Mine at Hutchinson, Kansas and its stockpiling of currency in an underground vault near Culpepper, Virginia.[18]

Not all risks, though, require elaborate, expensive defenses. More common recovery approaches to data base security include duplicate master files in a secure vault or confidential remote location.[19] Magnetic tape, the customary retention method, is quite inexpensive.

A recovery strategy is but one possible defense against teenage "crash clubs" or saboteurs working for a foreign government or private firm that would benefit from either unauthorized access to confidential information or undetected planting of erroneous data. A full security program would entail an array of prevention strategies. Preventing computer abuse involves two distinct issues: privacy protection against unauthorized retrieval and integrity protection against unauthorized tampering.[20] Passwords can restrict users to specific files and in some cases to only reading these files, not writing or altering them.[21] Even so, more nefarious intruders might penetrate password-protected systems with a *masquerade,* that is, by tapping a legitimate user's connection and later using his identification code and password to gain access to his files and the system. Linde lists 18 different kinds of attack against a computer's operating system.[22] Protection is particularly difficult with a distributed data base, less so with a single-location system.[23] A self-contained minicomputer without telecommunications service requires little more than physical security. Personnel security is important, as is high morale, which may encourage workers to be more vigilant in reporting possible breaches and in recommending improved procedures.[24] Effective security must be sufficiently comprehensive to protect against a wide variety of risks, including fire and inadvertent mistakes.[25]

Management's concern for data base security ought not become a paranoid obsession, preoccupied with nightmares of enemy spies, bent computer science majors, or disgruntled employees. Several discussions the author has had with persons overseeing system security have identified such commonplace occurrences as poor system design, lax hardware maintenance, inadequate air conditioning, poorly documented software, lax employee supervision, and inadequate user training as phenomena most likely to undermine the cost-effective and reliable performance of a computer data base. Indeed, air conditioning failures, disc crashes, misinterpretation of results, and inadvertent erasing of files probably destroy more data and waste more resources in a year than tamperers and terrorists do in a decade. In this sense the user needs as much protection from internal misuse and mismanagement as from external penetration.

Preservation and the Historical Record

The Electronic Transition also presents to historians of cartography the challenge of safeguarding mapping's past. Time and the environment threaten maps and mappable data with a slow yet determined force. Preserving for posterity the decaying paper medium of the recent cartographic record is exceeded in magnitude if not complexity by the task of developing safe, organized archives in which map data and mapping software will not be lost to static electricity, brittle plastics, and flaking coatings of iron oxide. Digital maps are by no means immune from aging and decay.

The ultimate enemy of the paper map is not its sometime competitor, the digital map, but the acid content of the paper itself. Paper is made by laying a mass of wet fibers on a porous frame, flattening the fibers, and draining away the water. Macerated cloth provided the fibers for most early papers, which had a high rag content. The rags were beaten into a pulp that could easily be spread in a thin sheet over the frame. To provide a harder writing surface, resistant to ink bleeding, the paper was then "sized" by dipping it into a warm gelatin. When rags became scarce in the late nineteenth century, makers of newsprint began to use wood pulp, which was then and is still far more abundant. Unfortunately, wood pulp retains *lignin,* which holds the short wood fibers together within the tree. Lignin breaks down into acids that cause the paper to turn yellow and flake apart. Paper also suffers from handling, humidity, light, heat, air pollution, and insects.[26]

Newspapers have used paper made with wood pulp as early as 1868, and there is a long history of attempts by libraries to deal with the deterioration of paper.[27] Most papers used in printing are susceptible to acid decay, and like librarians in general, map librarians have been concerned with deacidification and other preservation techniques.[28] Because they are not bound, map sheets are somewhat easier to treat with a solution that neutralizes, dissolves, and removes oxidizable material that might produce acid.[29] Yet caution is needed, as some treatments, such as lamination with adhesives, can precipitate other kinds of deterioration. Polyester film encapsulation, which avoids oxidation as well as the ill-effects of adhesives and lamination, is particularly promising for flat, unbound materials such as maps.[30] Photographic preservation and optical scanning are other approaches to conserving a map's information content and most of its design aesthetics.

Digital maps can be equally vulnerable to decay, so much so that several tape backup copies are routinely made even for ephemeral data sets developed only as intermediate steps in a research project. Tape can break, and a faulty write-head can produce meaningless tapes. If carried near a locally intense magnetic field, for example, in the vicinity of some aircraft engines, a tape's pattern of binary digits can be rendered partly or totally incomprehensible. The environment, aging, and wear all extract a toll.[31] Videodisc or microfilm storage is more durable, but over centuries some loss is inevitable unless new copies are made. In the long run nothing is permanent.

The ephemeral, ever-changing content of some digital maps presents another problem—how often to generate a copy for the historical archives. Should snapshots be taken at fixed time intervals or only after significant changes have occurred? And on what basis are the accumulated updates to be considered a significant change? Equally important is documentation of the origin of these changes. An archived digital map, like an alleged painting by an Old Master, is worth little without a reliable provenance.

Historians of cartography, of course, have interests wider than the artifactual record of maps and map data. Their concern with past manual techniques of surveying and engraving will evolve to include studies of present-day and future display hardware and program software. The increasingly complex institutions that adapt technology and manage its operation deserve a fuller treatment than survey organizations and map publishers have received in the past. Because the recent past and near future should leave a long-lasting imprint on mapping and geography, cartographic historians must develop the respect for the present of the social historian, the diplomatic historian, and the economic historian. Indeed, much useful information can best be collected now, through interviews and participant observation, as well as through more traditional archival and bibliographic channels. If cartography is to learn from its past, its historians must record and interpret its present.

Humanistic Challenges

The Electronic Transition presents educators and their institutions with a challenge almost as massive as that hurled at federal bureaucrats. In addition to expanding people's abilities to prepare their food, get to work, balance their checkbooks, and entertain themselves, modern technology will also increase the supply of map information available to the average citizen. More important, technology will not only generate more maps but also make it easier to make one's own maps. With personal computers, after all, the map need no longer be a rigid, unyielding icon. More noncartographers than ever can be their own mapmaker, their own Rand McNally. But will they have the skills and knowledge to do the job effectively? The professional cartographer might become more an educator and less a designer and manufacturer.

Lest this forecast of a huge increase in the number of do-it-yourself mapmakers seem like overly enthusiastic speculation, consider the following related predictions of several authorities on the future uses of computers and telecommunications. IBM Vice President Lewis Branscomb suggests that it may eventually be less expensive to regenerate a street map from satellite data than to construct one from municipal records.[32] Research scientist Joseph Martino, associate editor of the journal *Technological Forecasting and Social Change,* foresees by the year 2000 electronic libraries delivering to the home facsimile copies of books and other graphic materials.[33] Maps would surely be included. Political science professor Ted Becker predicts "electronic town meetings" through two-way cable television and the increased participation in government of a better informed citizenry.[34] Video maps would obviously be important, particularly in matters of zoning, regional planning, and redistricting. Hollis Vail, Audio Chairman of the World Future Society, sees the 1980s as the decade during which the computer terminal emerges as a key household appliance.[35] Widespread terminals would provide an outlet for such electronic mail as specially requested map data, to be manipulated and displayed at home by the user. Henry Freeman, director of research for the Policy Studies Corporation and trends editor of *American Printer* magazine, believes that desk-top mini-CAMIS (Computer Assisted Makeup Imaging System) units will promote the timely delivery of documents and allow authors to become their own publishers.[36] Although videotex might enable some map authors to publish their work even without in-the-home on-demand printers, high-resolution hardcopy units would extend the range of map products that a small self-publishing cartographer could offer clients. Alvin Toffler, author and undisputed dean of futurist journalism, predicts a shift of jobs, on a permanent or part-time basis, from the centralized office or manufacturing plant to the worker's "electronic cottage."[37] With high-capacity information networks and efficient graphics terminals, planners and other map-using professionals can readily consult with clients around the world without leaving home. And Robert Hamrin, a senior policy economist in the U.S. Environmental Protection Agency, notes that even by the 1970s information handling employed more people than mining, agriculture, manufacturing, and personal services combined. Hamrin foresees a significant reduction in assembly-line jobs and an increased use of robots.[38] Educators training students to become map drafters are promoting a short-lived skill.

The challenge for cartographic educators is two-fold: to shift instructional offerings for cartographers-in-training from readily obsolete manual skills to more long-lived topics in applied technology, management, geographic analysis, and aesthetics; and to expand training in map appreciation and map use at the introductory levels. Basic instruction in map use belongs as much in the elementary curriculum as at the high-school and college levels. Although by no means a traditional approach, instruction in map use might be taught formally and systematically outside geography, perhaps combined with related topics in computer science or graphic analysis. Perhaps with the increasing popularity of computer graphics, educational policy makers will finally recognize "graphicacy" as an essential skill for the educated, informed, productive citizen. And facility with maps is crucial to graphicacy.

Academic geography, the haven for most cartographic instruction, has yet another challenge—a research paradigm that recognizes mapping policy as an object of study that is as important to geography's own professional interests as it is to society at large. After all, given its ties to urban planning, regional development, and environmental protection, geography is at least in part a policy science. It is also an information science, with roots deep in exploration and with a name that means, quite literally, "earth description." In many ways it is unfortunate that geography's descriptive-inventory function was the hapless baby discarded ruthlessly with other methodological bathwater during the excesses of the Quantitative Revolution of the 1960s. Knowledge about places and the environment is significant, and much basic exploration and theorizing remains to be done. Now, because of high technology, geographers can prepare more complete and more effective descriptions and interpretations than before. Digital cartography provides a convenient mechanism not only for more relevant and informative maps, but also for the closer integration of maps and their associated written text.[39] To promote both its policy role and its journalistic role, geography needs a research orientation focused on mapping policy, information management, and related theoretical and technological concerns.[40]

Along another research frontier, geographic cartographers concerned with psychology might add vigilance and fatigue to their interests in perception and cognition. Map analysis with a VDU is even more error-prone than with paper maps, and periodic work breaks are necessary.[41] More ominous are such possible health hazards as face rash, backache, cataracts, and birth defects.[42] These concerns will surely affect the applications of and attitudes toward digital maps.

Aesthetics and communications are a final humanistic challenge for cartography's Electronic Transition. Maps should communicate and they should also please.[43] Like movies, plays, novels, TV documentaries, good children's television programs, and even magazine ads, the map is both a communication medium and an art object. It can convey a message as well as evoke interest and appreciation. Good design is as necessary in digital cartography as in any other endeavor yielding visual images, and good design must extend to system design and mapping software as well as to graphic symbols. Despite lingering memories of some hardly excusable bad maps produced in the early days of computer-assisted cartography, the computer can be a facilitator of style and taste as much as a perpetrator of haste and expediency. In extending the graphic artist's ability to apply color, time, motion, and the third dimension, the computer should greatly stimulate aesthetic creativity.[44] Digital cartography can produce refreshingly original and visually sound maps with meaningful consistency if an educated, sensitized public demands them.

NOTES
BIBLIOGRAPHY
GLOSSARY
INDEX

Notes

CHAPTER 1: Introduction

1 Commission III—Computer-assisted Cartography, International Cartographic Association, *A Glossary of Technical Terms in Computer-assisted Cartography* (Falls Church, Va.: American Congress on Surveying and Mapping, for the International Cartographic Association, 1980), pp. 17–18.

2 Arthur Robinson, Randall Sale, and Joel Morrison, *Elements of Cartography,* 4th ed. (New York: John Wiley and Sons, 1978), p. 276. A strong argument can be made for introducing two new terms: *real map* to describe traditional, hardcopy, analog maps and *virtual map* to describe both ephemeral images on a video screen and digital cartographic data files. There are three types of virtual map, differentiated according to the ease with which they may be manipulated by computer. See Harold Moellering, "Designing Interactive Cartographic Systems Using the Concepts of Real and Virtual Maps." In *Proceedings of the International Symposium on Computer-Assisted Cartography, Auto-Carto VI, October 16–21, 1983* (Ottawa: Steering Committee of Auto-Carto Six, 1983), vol. II, pp. 53–64.

3 F. W. Beers, G. P. Sanford, and others, *Atlas of Bennington County, Vermont* (New York: F. W. Beers, 1869).

4 Norman J. W. Thrower, *Maps and Man: An Examination of Cartography in Relation to Culture and Civilization* (Englewood Cliffs, N.J.: Prentice-Hall, 1972), pp. 102–5.

5 John L. Place, "The Land Use and Land Cover Map and Data Program of the U.S. Geological Survey: An Overview," *Remote Sensing of the Electromagnetic Spectrum* 4, no. 4 (October 1977): 1–9.

6 William B. Mitchell and others, *GIRAS: A Geographic Information Retrieval and Analysis System for Handling Land Use and Land Cover Data* (Washington: U.S. Government Printing Office, U.S. Geological Survey Professional Paper no. 1059, 1977).

7 Mark S. Monmonier, "Private-sector Mapping of Pennsylvania: A Selective Cartographic History for 1870 to 1974," *Proceedings of the Pennsylvania Academy of Science* 55 (1981): 69–74.

8 Arthur H. Robinson, *Early Thematic Mapping in the History of Cartography* (Chicago: University of Chicago Press, 1982).

9 Erwin Raisz, "Charts of Historical Cartography," *Imago Mundi* 2 (1937): 9–16.

10 Chauncey Starr and Richard Rudman, "Parameters of Technological Growth," *Science* 182 (1973): 358–64.

11 John H. Lienhard, "The Rate of Technological Improvement Before and After the 1830s," *Technology and Culture* 20 (1979): 515–30.

12 D. S. L. Cardwell, *Turning Points in Western Technology* (New York: Science History Publications, 1972), pp. 215–17.

13 M. J. Blakemore and J. B. Harley, *Concepts in the History of Cartography: A Review and Perspective* (Toronto: University of Toronto Press, 1980; also *Cartographica,* monograph no. 26).

14 See, for example, Jerome Dobson, "Automated Geography," *Professional Geographer* 35 (1983): 135–43.

CHAPTER 2: Location and Navigation

1 As an example, in older sections of Baltimore, Maryland, the street grid deviates by about 3 degrees from the true cardinal directions. In the American Midwest and West, street grids are commonly aligned to one of 34 principal meridians and base lines of the National Land System. In cities well removed from one of the more accurately surveyed *standard parallels* of this rectangular survey system, the street grid can also deviate noticeably from local meridians and parallels.

2 Norman J. W. Thrower, *Maps and Man* (Englewood Cliffs, N.J.: Prentice-Hall, 1972), pp. 5–8.

3 William Davenport, "Marshall Islands Navigation Charts," *Imago Mundi* 15 (1960): 19–26.

4 Ibid., p. 22.

5 Nathaniel Bowditch, *American Practical Navigator: An Epitome of Navigation,* corrected print (Washington: U.S. Government Printing Office, 1966), p. 20.

6 James Burke, *Connections* (Boston: Little, Brown and Co., 1978), p. 27.

7 Ibid., pp. 27–28.

8 Lloyd A. Brown, *The Story of Maps* (Boston: Little, Brown and Co., 1949), pp. 128–32.

9 Burke, *Connections,* p. 28.

10 Bowditch, *American Practical Navigator,* pp. 29–30.

11 Brown, *Story of Maps,* pp. 136–38.

12 Bowditch, *American Practical Navigator,* p. 18.

13 Brown, *Story of Maps,* pp. 67–69.

14 Bowditch, *American Practical Navigator,* p. 43.

15 Ibid.

16 Brown, *Story of Maps,* p. 206.

17 Derek Howse, *Greenwich Time and the Discovery of the Longitude* (Oxford: Oxford University Press, 1980), p. 164.

18 "Equation of Time," in *Encyclopedia Britannica,* 11th ed. (1910); and Howse, *Greenwich Time,* pp. 37–38.

19 Howse, *Greenwich Time,* pp. 194–220.

20 Bowditch, *American Practical Navigator,* pp. 51–53.

21 Howse, *Greenwich Time,* p. 131.

22 John D. Bossler, "A Note on Global Positioning System Activities," *Bulletin of the American Congress on Surveying and Mapping* no. 74 (1981): 39–40.

23 G. H. Rowe, "The Doppler Satellite Positioning Technique," *New Zealand Surveyor* 29 (1981): 608–24.

24 Larry D. Hothem, William E. Strange, and Madeline White, "Doppler Satellite Surveying System," *Journal of the Surveying and Mapping Division,* American Society of Civil Engineers 104 (1978): 79–91.

25 Rowe, "Doppler Satellite Positioning," p. 613; and Bossler, "Note on Global Positioning," p. 39.

26 Robert J. Urick, *Principles of Underwater Sound for Engineers,* 2nd ed. (New York: McGraw-Hill, 1975), pp. 1–15.

27 G. D. Dunlap and H. H. Shufeldt, *Dutton's Navigation and Piloting,* 12th ed. (Annapolis, Md.: U.S. Naval Institute, 1969), pp. 306–19; and David H. Gray, "The Preparation of Loran-C Lattices for Canadian Charts," *Canadian Surveyor* 34 (1980): 277–95.

28 Daniel A. Panshin, "What You Should Know about Loran-C Receivers," Marine Electronics Series, Oregon State University, Extension Marine Advisory Program, April 1979.

29 Kosta Tsipis, "Cruise Missiles," *Scientific American* 236, no. 2 (February 1977): 20–29.

30 Ibid., p. 24.

31 Ibid., p. 23.

32 D. E. Richardson, "The Cruise Missile: a Strategic Weapon for the 1980s," *Electronics and Power* 23 (1977): 896–901; and Tsipis, "Cruise Missiles," p. 23.

33 Tsipis, "Cruise Missiles," pp. 22–24.
34 Ibid., p. 29.
35 V. David Hopkin and Robert M. Taylor, *Human Factors in the Design and Evaluation of Aviation Maps* (Neuilly Sur Seine: NATO, Advisory Group for Aerospace Research and Development, no. AGARD–AG–225, 1979), p. 48.
36 Ibid., pp. 49–54.
37 Ibid., pp. 31–41.
38 The epigram at the front of John Andrew's history of British mapping in Ireland is a telling quote credited to Lord Salisbury in 1883: "The most disagreeable part of the three kingdoms is Ireland, and therefore Ireland has a splendid map." See J. H. Andrews, *A Paper Landscape: The Ordnance Survey in Nineteenth-century Ireland* (Oxford: Oxford University Press, 1975), p. v.
39 The Department of Defense will encrypt the signal from GPS satellites so that civilians with a single receiver can estimate absolute location only to the nearest 500 m [1,600 ft]. With two or more receivers simultaneously receiving a signal from the same satellite, though, relative position may be estimated to within several centimeters. For a discussion of the capabilities for land surveyors of this all-weather system, expected to be in operation by 1990, see Adam Chrzanowski, and others, "A Forecast of the Impact of GPS on Surveying," *Technical Papers of the American Congress on Surveying and Mapping* 43rd Annual Meeting, 1983, pp. 625–34. For discussion of the use of GPS for marine navigation, see Bradley O. Montgomery, "The NAVSTAR Global Positioning System," *Professional Surveyor* 3, no. 5 (September/October 1983): 13–17.

CHAPTER 3: Boundaries and Surveys

1 See, for example, Russell C. Brinker and Paul R. Wolf, *Elementary Surveying,* 6th ed. (New York: IEP—A Dun-Donnelley Publisher, 1977), pp. 5–9; and Milton O. Schmidt and William Horace Rayner, *Fundamentals of Surveying,* 2nd ed. (New York: D. Van Nostrand Co., 1978), pp. 3–7.
2 John G. McEntyre, *Land Survey Systems* (New York: John Wiley and Sons, 1978), p. 4.
3 Schmidt and Rayner, *Fundamentals of Surveying,* pp. 4–6.
4 James Burke, *Connections* (Boston: Little, Brown and Co., 1978), pp. 253–58.
5 Ibid., p. 258.

6 Ibid., pp. 258–60.
7 David Eugene Smith, *History of Mathematics, Volume II: Special Topics of Elementary Mathematics* (New York: Ginn and Co., 1925; New York: Dover Publications, 1958), pp. 346–47.
8 Don W. Thompson, *Men and Meridians, Volume I: Prior to 1867* (Ottawa: Information Canada, 1966), p. 21.
9 Ibid., pp. 21, 24.
10 Smith, *History of Mathematics, Vol. II*, p. 627.
11 The utility of logarithms is expressed in a tale about Noah and the ark. As those familiar with the Biblical account will recall, God commanded Noah to build an ark and to gather into it a male and female of each species. Noah obeyed. God then made it rain for 40 days and 40 nights and thus drowned or otherwise disposed of "the wicked" as well as most of the animal population. When the flood subsided, God commanded Noah to open the ark and release the animals. Noah eagerly complied. God then commanded the animals to go forth, and increase and multiply. This the animals were only too glad to do, except for an unfortunate pair of snakes—adders, to be precise—which were having great difficulties in fulfilling God's command. Noah, anxious that God's wrath would not again be incurred, pondered this problem fitfully for several days before commanding his sons to cut down a tall tree, saw it into logs, and build a table. Noah then invited the snakes onto the table, and they were, with glee and vigor, able to fulfill God's directive. The moral of this tale is, of course, that with the aid of a log table even adders can multiply. Arithmetically, a product can be found by adding the logarithms of the factors and then taking the antilog of this sum. For computational formulas useful in triangulation see, for example, George D. Whitmore, *Advanced Surveying and Mapping* (Scranton, Pa.: International Textbook Co., 1949), pp. 94–209.
12 Michael V. Smirnoff, *Measurements for Engineering and Other Surveys* (Englewood Cliffs, N.J.: Prentice-Hall, 1962), pp. 87–94.
13 Ibid., pp. 84–85.
14 Ibid., pp. 83, 87.
15 Burke, *Connections*, pp. 268–73; and J. H. Andrews, *A Paper Landscape: The Ordnance Survey in Nineteenth-century Ireland* (Oxford: Oxford University Press, 1975), p. 42.
16 K. D. Froome and L. Essen, *The Velocity of Light and Radio Waves* (London and New York: Academic Press, 1969), pp. 114–27; John R. Greene, "Accuracy Evaluation in Electro-Optical Distance-Measuring Instruments," *Surveying and Mapping* 37 (1977): 247–

56; and Simo Laurila, *Electronic Surveying and Mapping* (Columbus: Ohio State University, Institute of Geodesy, Photogrammetry and Cartography, Publication no. 11, 1960), pp. 186–203.

17 John Noble Wilford, *The Mapmakers* (New York: Alfred A. Knopf, 1981), pp. 306–8.

18 These figures are for a hypothetical, composite instrument based on data reported in Timothy P. Bell, "A Practical Approach to Electronic Distance Measurement," *Surveying and Mapping* 38 (1978): 335–41.

19 Paul R. Wolf and Steven D. Johnson, "Trilateration with Short Range EDM Equipment and Comparison with Triangulation," *Surveying and Mapping* 34 (1974): 337–46. Also see David F. Mezera, "Trilateration Adjustment Using Unit Corrections Derived from Least Squares," *Surveying and Mapping* 43 (1983): 315–29.

20 See, for example, R. Hirvonen, *Adjustment by Least Squares in Geodesy and Photogrammetry* (New York: Frederick Ungar Publishing Co., 1971); David Clark, *Plane and Geodetic Surveying, Volume Two: Higher Surveying* (London: Constable, 1973), pp. 102–67; and Atef A. Elassal, "Generalized Adjustment in Least Squares (GALS)," *Photogrammetric Engineering and Remote Sensing* 49 (1983): 201–6.

21 Hirvonen, *Adjustment by Least Squares*, pp. 1–10.

22 Urho A. Uotila, "Useful Statistics for Land Surveyors," *Surveying and Mapping* 33 (1973): 67–77.

23 Erwin Raisz, *Mapping the World* (London: Abelard-Schuman, 1956), pp. 59–63; and Josef V. Konvitz, "Redating and Rethinking the Cassini Geodetic Surveys of France, 1730–1750," *Cartographica* 19, no. 1 (Spring 1982): 1–15.

24 D. H. Maling, *Coordinate Systems and Map Projections* (London: George Philip and Son Limited, 1973), pp. 3–17; and Irene Fischer, "Is the Astrogeodetic Approach in Geodesy Obsolete?" *Surveying and Mapping* 34 (1974): 121–30.

25 John D. Bossler, "New Adjustment of North American Datum," *Journal of the Surveying and Mapping Division, Proceedings of the American Society of Civil Engineers* 108, no. SU2 (August 1982): 47–52. The adjustment involved solving a system of 400,000 equations in 400,000 unknowns. See Gina Bari Kolata, "Geodesy: Dealing with an Enormous Computer Task," *Science* 200 (1978): 421–22, 466; and J. E. Colcord, "The Surveying Engineer and NAD–83," *Journal of the Surveying and Mapping Division, American Society of Civil Engineers* 107, no. SU1 (November 1981): 25–31.

26 H. R. Lippold, Jr., "Readjustment of the National Geodetic Vertical Datum," *Surveying and Mapping* 40 (1980): 155–64.

27 A. Bannister and S. Raymond, *Surveying,* 4th ed. (London: Pitman Publishing Limited, 1977), pp. 265–80.

28 The time required for stereoplotting will depend upon map scale, contour interval, terrain roughness, and extent of urban development, among other factors. See John E. Combs and others, "Planning and Executing the Photogrammetric Project." In *Manual of Photogrammetry,* 4th ed. (Falls Church, Va.: American Society of Photogrammetry, 1980), pp. 367–412.

29 See Morris M. Thompson and Heinz Gruner, "Foundations of Photogrammetry." In *Manual of Photogrammetry,* 4th ed. (Falls Church, Va.: American Society of Photogrammetry, 1980), pp. 1–36; Russell K. Bean, "The Orthophotoscope and Its Development," *Canadian Surveyor* 2 (1968): 38–45; and Dierk Hobbie and Hans W. Faust, "Z-2 ORTHOCOMP, the New High Performance Orthophoto Equipment from Zeiss," *Photogrammetric Engineering and Remote Sensing* 49 (1983): 635–40.

30 See Frederick J. Doyle, "Digital Terrain Models: An Overview," *Photogrammetric Engineering and Remote Sensing* 44 (1978): 1481–84; and Pinhas Yoeli, "Digital Terrain Models and Their Cartographic and Cartometric Utilisation," *Cartographic Journal* 20 (1983): 17–22.

31 Norman J. W. Thrower and John R. Jensen, "The Orthophoto and Orthophotomap: Characteristics, Development and Application," *American Cartographer* 3 (1976): 39–56.

32 The precision with which contours might be drawn depends upon the accuracy of the ground control, the precision of the stereoplotting equipment, flying conditions, lens distortion, and flight height. A convenient rule of thumb estimates contour accuracy as $\pm 3H/10,000$, where H is flight height in meters. See Bannister and Raymond, *Surveying,* p. 584.

33 See S. Jack Friedman and others, "Automation of the Photogrammetric Process." In *Manual of Photogrammetry,* 4th ed. (Falls Church, Va.: American Society of Photogrammetry, 1980), pp. 699–722.

34 Walter H. Treftz, "An Introduction to Inertial Positioning As Applied to Control and Land Surveying," *Surveying and Mapping* 41 (1981): 59–67.

35 William H. Chapman, "Proposed Specifications for Inertial Surveying," *Technical Papers of the American Congress on Surveying and Mapping* 43rd Annual Meeting, 1983, pp. 287–93.

36 Treftz, "An Introduction to Inertial Positioning," p. 67.
37 For a discussion of British thinking, at a particularly cost-conscious time, on the proper role of government in mapping, see Gerald McGrath, "Re-defining the Role of Government in Surveys and Mapping/A View of Events in the United Kingdom," *Cartographica* 19, nos. 3/4 (Autumn/Winter 1982): 44–52.

CHAPTER 4: Aerial Reconnaissance and Land Cover Inventories

1 See Hildegard Binder Johnson, *Order Upon the Land: The U.S. Rectangular Land Survey and the Upper Mississippi Country* (New York: Oxford University Press, 1976), pp. 143–48.
2 Norman J. W. Thrower, "Cadastral Survey and County Atlases of the United States," *Cartographic Journal* 9 (1972): 43–51.
3 Atlas publishing, which could be highly profitable, often involved highly questionable business practices. A book on consumer frauds, published in 1890, devoted 67 of its 300 pages to county and state atlases and wall maps, which shared the limelight with lightening rod agents, fruit tree swindles, patent medicines, and other unsavory enterprises. See *How 'Tis Done: A Thorough Ventilation of the Numerous Schemes Conducted by Wandering Canvassers Together with the Various Advertising Dodges for the Swindling of the Public* (Syracuse: W. I. Pattison, 1890).
4 Henry Gannett, "The Mother Maps of the United States," *National Geographic Magazine* 4 (1892): 101–16.
5 Ibid., pp. 101–2.
6 Ibid., p. 111.
7 Henry Gannett, "The Mapping of the United States," *Scottish Geographical Magazine* 8 (1892): 150–53. Percentage of area mapped is based on the present 7.7 million square km of the original, "lower" 48 states.
8 Printing and distribution are normally provided by the federal government. See Melvin Y. Ellis, ed., *Coastal Mapping Handbook* (Washington: U.S. Government Printing Office, 1978), p. 15–16.
9 Mark S. Monmonier, "Topographic Map Coverage of Pennsylvania: A Study in Cartographic Evolution," *Proceedings of the Pennsylvania Academy of Science* 56 (1982): 61–66.
10 For a philosophical discussion of the distinction between general and thematic maps, see Arthur H. Robinson and Barbara Bartz Petchenik, *The Nature of Maps* (Chicago: University of Chicago Press, 1976), pp. 116–21.

11 See Robert S. Quackenbush, Jr., Arthur C. Lundahl, and Edward Monsour, "Development of Photo Interpretation." In *Manual of Photographic Interpretation* (Washington: American Society of Photogrammetry, 1960), pp. 1–18.

12 William A. Fischer and others, "History of Remote Sensing." In *Manual of Remote Sensing,* vol. I (Falls Church, Va.: American Society of Photogrammetry, 1975), pp. 27–50 [ref. on pp. 27–28].

13 Quackenbush, Lundahl, and Monsour, "Development of Photo Interpretation," p. 5.

14 Harland Manchester, *New Trail Blazers of Technology* (New York: Charles Scribner's Sons, 1978), pp. 131–41.

15 For a discussion of German preparations during the late 1930s for the push to the Atlantic and the invasion of Britain, see Fischer and others, "History of Remote Sensing," pp. 32–33. Britain, forced to rely upon aerial reconnaissance of the European mainland after its defeat at Dunkirk in 1940, detected German preparations for the invasion and launched an air attack that thwarted Nazi plans. For an interesting account of British use of photoreconnaissance, see Constance Babington-Smith, *Air Spy* (New York: Harper and Row, 1957).

16 See George W. Goddard and DeWitt S. Copp, *Overview: A Lifelong Adventure in Aerial Photography* (New York: Doubleday, 1969). Goddard's major contribution—and one accepted more readily by the Navy than by his own branch, the Air Force—was the continuous strip camera. The camera was able to record sharp, detailed images from a rapidly moving reconnaissance aircraft. In 1962 the strip camera was important in detecting, documenting, and eventually verifying the removal of Soviet missiles in Cuba.

17 Quackenbush, Lundahl, and Monsour, "Development of Photo Interpretation," pp. 12–13.

18 See J. Ronald Eyton, "Landsat Multitemporal Color Composites," *Photogrammetric Engineering and Remote Sensing* 49 (1983): 231–35; and Thomas M. Lillesand and Ralph W. Kiefer, *Remote Sensing and Image Interpretation* (New York: John Wiley and Sons, 1979), pp. 54–58.

19 The RBV imagery was of higher cartographic quality than that of the Landsat multispectral scanner system, but because of equipment problems, comparatively few RBV scenes were recorded in the early years of Landsat. See L. Ralph Baker and others, "Electro-Optical Remote Sensors with Related Optical Sensors." In *Manual of Remote Sensing* (Falls Church, Va.: American Society of Photogrammetry, 1975), pp. 326–66.

20 Band 7, the wider of the two reflected infrared bands, was recorded with a range of 0 to 63. See Donald S. Lowe and others, "Imaging and Nonimaging Sensors," In *Manual of Remote Sensing* (Falls Church, Va.: American Society of Photogrammetry, 1975), pp. 367–97.

21 Although the grid of rectangular pixels is a convenient concept for organizing the data, the grid format reflects only the appropriate arrangement of the pixels. Each pixel records the reflectance from a ground cell about 79 m on a side, but these "spots" are not perfect squares. Moreover, because the Earth is rotating beneath the satellite, successive scan lines are offset slightly so that the columns of the grid are not truly perpendicular to the rows of scan lines. Other geometric distortions result from variations in the speed of the oscillating mirror, variation along the scan line in distance between ground and sensor, and other factors. See Alden P. Colvocoresses, "Applications to Cartography: Introduction," *ERTS–1, A New Window on Our Planet* (Washington: U.S. Government Printing Office, U.S. Geological Survey Professional Paper no. 929, 1976), pp. 12–22.

22 F. J. Doyle, "Satellite Systems for Cartography," *ITC Journal* no. 1981–2 (1981): 153–70. For discussion of a modified Mapsat design, see Alden P. Colvocoresses, "The Relationship of Acquisition Systems to Automated Stereo Correlation," *Photogrammetric Engineering and Remote Sensing* 49 (1983): 539–44. Stereoscopic viewing will be possible with imagery from the SPOT satellite designed by and built for the French Centre National d'Etudes Spatiales.

23 For further examples see A. H. Benny and G. J. Dawson, "Satellite Imagery as an Aid to Bathymetric Charting in the Red Sea," *Cartographic Journal* 20 (1983): 5–16; John R. Jensen, "Biophysical Remote Sensing," *Annals of the Association of American Geographers* 73 (1983): 111–32; and Thomas M. Lillesand and Ralph W. Kiefer, *Remote Sensing and Image Interpretation*, pp. 543–53.

24 Frederick J. Doyle, "The Next Decade of Satellite Remote Sensing," *Photogrammetric Engineering and Remote Sensing* 44 (1978): 155–64.

25 See Homer Jensen and others, "Side-looking Airborne Radar," *Scientific American* 237, no. 4 (October 1977): 84–95.

26 See, for example, Fawwaz T. Ulaby, Gerald A. Bradley, and Myron C. Dobson, "Microwave Backscatter Dependence on Surface Roughness, Soil Moisture, and Soil Texture: Part II—Vegetation-Covered Soils," *IEEE Transactions on Geoscience Electronics* GE–17 (1979): 33–40.

27 Manouher Naraghi, William Stromberg, and Mike Daily, "Geometric Rectification of Radar Imagery Using Digital Elevation Models," *Photogrammetric Engineering and Remote Sensing* 49 (1983): 195–99; and Sherman S. C. Wu, "Geometric Corrections of Side-looking Radar Images," *Technical Papers of the American Society of Photogrammetry* 49th Annual Meeting, 1983, pp. 354–64.

28 See Charles Elachi, "Spaceborne Imaging Radar: Geologic and Oceanographic Applications," *Science* 209 (1980): 1073–82.

29 Alden P. Colvocoresses and others, "Platforms for Remote Sensors." In *Manual of Remote Sensing* (Falls Church, Va.: American Society of Photogrammetry, 1975), pp. 539–88; and Eric C. Barnett and Michael G. Hamilton, "The Use of Geostationary Satellite Data in Environmental Science," *Progress in Physical Geography* 6 (1982): 159–214.

30 James V. Taranik and Mark Settle, "Space Shuttle: A New Era in Terrestrial Remote Sensing," *Science* 214 (1981): 619–26.

31 Specific details, of course, are classified TOP SECRET, but scuttlebutt abounds in generally reliable sources. See, for example, D. E. Richardson, "Spy Satellites: Somebody Could Be Watching You," *Electronics and Power* 24 (1978): 573–76.

32 John R. G. Townshend, "The Spatial Resolving Power of Earth Resources Satellites," *Progress in Physical Geography* 5 (1981): 32–55.

33 See, for example, Johannes G. Moik, *Digital Processing of Remotely Sensed Images* (Washington: National Aeronautics and Space Administration, no. NASA SP–431, 1980), pp. 271–73.

34 See, for example, King Sun Fu and T. S. Yu, *Statistical Pattern Classification Using Contextual Information* (New York: Research Studies Press, a division of John Wiley and Sons, 1980); and Kamlesh Lulla, "The Landsat Satellites and Selected Aspects of Physical Geography," *Progress in Physical Geography* 7 (1983): 1–45.

35 Splitting a band may not always be desirable, particularly if the reflectance of objects of interest varies unpredicatably and uninformatively across the two new bands.

36 See John R. Jensen, "Urban Change Detection Mapping Using Landsat Data," *American Cartographer* 8 (1981): 127–47.

37 Orbit adjustments are needed every other day or so to compensate for atmospheric drag and the gravitational attraction of the Sun and Moon. The satellite's "mass expulsion subsystem" adjusts the orbit by emitting a measured amount of gas in a direction opposite that in which the satellite needs "a push" to retain the intended orbit. See

Landsat Data Users Handbook (Reston, Va.: U.S. Geological Survey, 1979), pp. 5–2 to 5–7. Also see Toyohisa Kaneko, "Evaluation of LANDSAT Image Registration Accuracy," *Photogrammetric Engineering and Remote Sensing* 42 (1976): 1285–99.

38 A classifier that examines neighboring reflectances may be useful for multispectral imagery of a single scene. Yet the grids for different bands might not correspond exactly even when recorded simultaneously. See, for example, Philip H. Swain, Howard Jay Siegel, and Bradley W. Smith, "Contextual Classification of Multispectral Remote Sensing Data Using a Multiprocessor System," *IEEE Transactions on Geoscience and Remote Sensing* GE–18 (1980): 197–203.

39 See, for example, David L. Milgram and Azriel Rosenfeld, "Object Detection in Infrared Images." In Leonard Bolc and Zenon Kulpa, eds., *Digital Image Processing Systems* (Berlin: Springer-Verlag, 1981), pp. 228–353; and R. Michael Hord, "Digital Enhancement of Landsat MSS Data for Mineral Exploration." In William L. Smith, ed., *Remote Sensing Applications for Mineral Exploration* (Stroudsburg, Pa.: Dowden, Hutchinson, and Ross, 1977), pp. 235–50.

40 U.S. Bureau of the Census, *Census of Agriculture, 1969*, Vol. 5, Special Reports, Part 15, Graphic Summary (Washington: U.S. Government Printing Office, 1973), pp. 14–15.

41 For discussion of plans by the United States government in the early 1980s to sell the nation's earth resources satellite system to a private firm or consortium, see M. Mitchell Waldrop, "What price Privatizing Landsat?" *Science* 219 (1983): 752–54. For an illustration of objections to the proposed sale, see Thomas M. Lillesand, "Issues Surrounding the Commercialization of Civil Land Remote Sensing from Space," *Photogrammetric Engineering and Remote Sensing* 49 (1983): 495–504.

42 Charles K. Paul and Adolfo C. Mascarenhas, "Remote Sensing in Development," *Science* 214 (1981): 139–45.

43 See, for example, Michel T. Halbouty, "Geologic Significance of Landsat Data for 15 Giant Oil and Gas Fields," *American Association of Petroleum Geologists Bulletin* 64 (1980): 8–36.

CHAPTER 5: Decision Support Systems

1 See G. R. Wagner, "Decision Support Systems: The Real Substance," *Interface* 11, no. 2 (April 1981): 77–86.

2 Hyman Alterman, *Counting People: The Census in History* (New York: Harcourt, Brace and World, 1969), pp. 17–28.

3 Fulmer Mood, "The Rise of Official Statistical Cartography in Aus-

tria, Prussia, and the United Statees, 1855–1872," *Agricultural History* 20 (1946): 209–25.

4 See U.S. Bureau of the Census, *Data Access Descriptions: Geography,* DAD No. 33, May 1979, pp. 6–8.

5 Roger A. Herriot, "The 1980 Census: Countdown for a Complete Count," *Monthly Labor Review* 102, no. 9 (September 1979): 3–13.

6 In 1980, about 10 percent of the rural population received a hand-delivered, self-administered questionnaire from their letter carrier. The remainder, largely in urban areas, received the mail-out and mail-back questionnaire. About 80 percent were expected to return their questionnaires without any prompting, and telephone and mail contacts were often tried before resorting to a personal visit. See Herriot, "The 1980 Census: Countdown for a Complete Count."

7 Other census areas such as the Enumeration District, the Block Group, and the State Economic Area are also used. For a more complete account, see U.S. Bureau of the Census, *Data Access Descriptions: Geography;* and Henry S. Shryock, Jacob S. Siegel and Associates, *The Methods and Materials of Demography* (Washington: U.S. Government Printing Office, 1971), pp. 113–133.

8 Marvin S. White, Jr., "A Geometrical Model for Error Detection and Correction." In *Proceedings of the International Symposium on Computer-Assisted Cartography, Auto-Carto III, January 16–20, 1978* (Falls Church, Va.: American Congress on Surveying and Mapping, 1979), pp. 439–56.

9 See U.S. Bureau of the Census, *Data Access Descriptions: Geography,* p. 13; and Morton A. Meyer, Frederick R. Broome, and Richard H. Schweitzer, Jr., "Color Statistical Mapping by the U.S. Bureau of the Census," *American Cartographer* 2 (1975): 100–117.

10 Such unstructured data as independent area polygons unrelated to their neighbors are termed "spaghetti files." See Thomas K. Peucker and Nicholas Chrisman, "Cartographic Data Structures," *American Cartographer* 2 (1975): 55–69.

11 The atlas also contained four larger-scale maps for each of seven selected subnational regions. See D. W. Rhind, I. S. Evans, and M. Visvalingam, "Making a National Atlas of Population by Computer," *Cartographic Journal* 17 (1980): 3–11.

12 See Olof Wastesson, "Computer Cartography and Geographic Information Systems in Sweden." In Olof Wastesson, Bengt Rystedt, and D. R. F. Taylor, eds., *Computer Cartography in Sweden* (Toronto: University of Toronto Press, 1977; also *Cartographica,* monograph no. 20), pp. 7–9.

13 Stephan W. Miller, "A Compact Raster Format for Handling Spatial

Data," *Technical Papers of the American Congress on Surveying and Mapping* Fall 1980, paper CD–4–A.

14 For a detailed discussion of CRT technology see Samuel Davis, *Computer Data Displays* (Englewood Cliffs, N.J.: Prentice-Hall, 1969), pp. 10–48; and Sol Sherr, *Electronic Displays* (New York: John Wiley and Sons, 1979), pp. 69–181.

15 Vector-mode CRTs can also plot lines in color, although these displays are seldom used for maps. See Sherr, *Electronic Displays*, pp. 420–22; and Turner Whitted, "Some Recent Advances in Computer Graphics." *Science* 215 (1982): 767–74.

16 Light in the three additive primary hues, blue, green, and red, can be combined to yield intermediate hues. For example, blue light added to red light produces magenta light, whereas adding green and reducing the intensity of the blue yields orange. See, for example, Sherr, *Electronic Displays*, pp. 16–19.

17 Sherr, *Electronic Displays*, pp. 212–21 and 512–24; and Donald K. Wedding and Roger E. Ernsthausen, "Large-Area Flat Panel Displays," *Computer Graphics World* 6, no. 4 (April 1983): 68–70.

18 See Robert C. Tsai, "High Data Density 4-color LCD System," *Information Display* no. 5–81 (May 1981): 3–6; and Sherr, *Electronic Displays*, pp. 186, 221–32.

19 For a straightforward introduction to the physical principles of holograms, see Winston E. Kock, *Lasers and Holography*, 2nd ed., enlarged (New York: Dover Publications, 1981). For examples of cartographic applications, see Geoffrey Dutton, "American Graph Fleeting: A Computer-Holographic Map Animation." In *Computer Mapping in Education, Research, and Medicine* (Cambridge, Mass.: Harvard University, Laboratory for Computer Graphics and Spatial Analysis, 1979), pp. 53–62; and James J. McGrath, "Contemporary Map Displays." In North Atlantic Treaty Organization, Advisory Group for Aerospace Resarch and Development, *Guidance and Control Displays*, AGARD Conference Proceedings No. 96, February 1972, pp. 13–1 to 13–16. See also J. W. Phillips, P. L. Ransom, and R. M. Singleton, "On the Construction of Holograms and Halftone Pictures with an Ink Plotter," *Computer Graphics and Image Processing* 4 (1975): 200–208.

20 See, for example, Bruce Eric Brown, "Computer Graphics for Large Scale Two- and Three-Dimensional Analysis of Complex Geometries," *Computer Graphics* 13, no. 2 (August 1979): 31–40; and John A. Roese, "Stereoscopic Computer Graphics for Simulation and Modeling," *Computer Graphics* 13, no. 2 (August 1979): 41–47. For a description of a system that uses a vibrating and deforming

mirror to produce a three-dimensional image, see Hank Stover, "Graphics System Displays True 3D Image," *Mini-Micro Systems* 14, no. 12 (December 1981): 121–23.

21 For a concise review of the physical principles of a variety of graphic display devices, see B. Kazan, "Materials Aspects of Display Devices," *Science* 208 (1980): 927–37.

22 Davis, *Computer Data Displays*, pp. 189–92; and Wolfgang K. Giloi, *Interactive Computer Graphics* (Englewood Cliffs, N.J.: Prentice-Hall, 1978), pp. 189–94.

23 Davis, *Computer Data Displays*, pp. 180–85; Giloi, *Interactive Computer Graphics*, pp. 194–97; and Vic Kley, "Pointing Device Communication," *Computer Graphics World* 6, no. 11 (November 1983): 69–72.

24 See Elias Prado, "Voice Input for CAD/CAM," *Computer Graphics World* 6, no. 6 (June 1983): 111–13.

25 See A. R. Boyle, "Development in Equipment and Techniques," *Progress in Contemporary Cartography* 1 (1980): 39–57.

26 Arthur Koestler, *The Act of Creation* (New York: Macmillan, 1964), pp. 35–45.

27 F. J. Ormeling, Sr., "The Purpose and Use of National Atlases." In Barbara J. Gutsell, ed., *The Purpose and Use of National Atlases* (Toronto: University of Toronto Press, 1979; also *Cartographica*, monograph no. 23), pp. 11–23.

28 Salichtchev's commission's report, *Atlas Nationaux*, was presented at Stockholm in 1960 to the 19th Congress of the International Geographical Union (IGU). A translated and edited version was published in English in 1972. See K. A. Salichtchev, ed., *National Atlases*, published as *Cartographica*, monograph no. 4, 1972.

29 Ormeling, "The Purpose and Use of National Atlases," p. 13.

30 See James R. Anderson, "The National Atlas of the United States." In Barbara J. Gutsell, ed., *The Purpose and Use of National Atlases* (Toronto: University of Toronto Press, 1979; also *Cartographica*, monograph no. 23), pp. 35–39. In the late 1970s, the U.S. Geological Survey had begun to plan for a second edition of the *National Atlas of the United States,* but by 1983 the project had not been funded. Planning included a computerized system for the development, review, and interactive management of the table of contents. See Mark S. Monmonier, "Automated Techniques in Support of Planning for the National Atlas," *American Cartographer* 8 (1981): 161–68.

31 Richard Groot, "Canada's National Atlas Program in the Computer Era." In Barbara J. Gutsell, ed., *The Purpose and Use of National*

Atlases (Toronto: University of Toronto Press, 1979; also *Cartographica,* monograph no. 23), pp. 41–52.

32 Heinrich Gutersohn, "Atlas der Schweiz—ein Rueckblick," *Geographica Helvetica* 34, no. 4 (1979): 181–88.

33 Groot, "Canada's National Atlas Program in the Computer Era," pp. 42–44.

34 For a general introduction to benefit-cost analysis, see Peter G. Sassone and William A. Schaffer, *Cost-Benefit Analysis: A Handbook* (New York: Academic Press, 1978). For water resource and other public works projects, the U.S. Army Corps of Engineers apparently has considered a ratio of unity sufficient, even when the ratio itself is inflated. See "Cost-benefit Trips Up the Corps," *Business Week,* Industrial Edition, no. 2573 (19 February 1979): 96–97. A benefit-cost ratio of 2 is generally considered highly favorable, especially if the estimate is conservative.

35 For an insightful commentary on many aspects of the Canadian atlas, see Henry W. Castner, "Concept Before Content? A Question in Atlas Design with Special Reference to the National Atlas of Canada," *Canadian Geographer* 20 (1976): 224–32.

36 N. L. Nicholson, "Review: *Atlas of Canada,*" *Cartographica* 18, no. 3 (Autumn 1981): 133–34.

37 Groot, "Canada's National Atlas Program in the Computer Era," p. 45.

38 The DIDS (Decision Information Display System) experiment of the early 1980s demonstrated the need for reviewing the gratuitous contributions of data by federal agencies. DIDS was an interactive DSS developed originally by NASA for the White House. It produced choropleth maps for the most part, for state units, counties, or metropolitan census tracts. Various federal agencies were required by the Office of Management and Budget to contribute from their budgets to the support of DIDS. These and other agencies also contributed data, which were included in the DIDS data base without external review. Numerous geographic distributions displayed by DIDS showed "missing data" for many areal units. Example maps promoting the system included choropleth maps with categories based on absolute counts rather than conceptually correct ratios or percentages. Unless the user deliberately changed them, highly contrasting hues, rather than an appropriate, progressive series of light-to-dark tones, represented value differences between choropleth categories. See Mark Stephen Monmonier, "DIDS—A Defacto National Atlas?," *Bulletin, Geography and Map Division, Special Libraries As-*

sociation no. 132 (June 1983): 2–7; and Edward K. Zimmermann, "The Evolution of the Domestic Information Display System: Toward a Government Public Information Network," *Review of Public Data Use* 8 (1980): 60–81.

39 See Harold Moellering, "The Challenge of Developing a Set of National Digital Cartographic Data Standards for the United States," *Technical Papers of the American Congress on Surveying and Mapping* 42nd Annual Meeting, 1982, pp. 201–12.

40 Robert L. Gignilliat, "Cleaning Up, Matching and Merging Data Files," *Review of Public Data Use* 4, no. 2 (March 1976): 9–15. Considerably more data are available on magnetic tape than appear in the printed reports of the Census of Population and Housing. See Martha Farnsworth Riche, "Choosing 1980 Census Data Products," *American Demographics* 3, no. 11 (December 1981): 12–16.

41 Malcolm J. Stephens, "The USGS 1:2,000,000–scale Digital Data Base," *Technical Papers of the American Congress on Surveying and Mapping* 40th Annual Meeting, 1980, pp. 436–43.

42 *Lab-Log 1980* (Cambridge, Mass.: Laboratory for Computer Graphics and Spatial Analysis, Harvard Graduate School for Design, 1980), pp. 28–29.

43 For a review of features available in turnkey systems for personal use, see Robert Perry, "Personal Computers as the EE's Best Friend," *Electronics* 54, no. 8 (21 April 1981): 115–60. For a comparison of turnkey and other approaches to developing a digital mapping system, see National Research Council, Panel on a Multipurpose Cadastre, *Procedures and Standards for a Multipurpose Cadastre* (Washington, D.C.: National Academy Press, 1982), pp. 50–54.

44 For an insightful discussion of the difficulties of coordinating large software projects, see Frederick P. Brooks, Jr., *The Mythical Man-Month: Essays on Software Engineering* (Reading, Mass.: Addison-Wesley Publishing Co., 1975).

45 For a concise guide to selecting a home computer, see Martin Webster, "Buying a Minicomputer," *Journal of Geography in Higher Education* 4 (1980): 42–50.

46 For a concise review of smart terminals, see George B. Bernstein and Arnold S. Kashar, *Intelligent Terminals: Functions, Specifications, and Applications* (Wellesley, Mass.: Q.E.D. Information Sciences, 1978).

47 For a review of built-in graphics functions available from a variety of software developers as well as a discussion of methodology and standards, see R. H. Ewald and R. Fryer, eds., "Final Report of the

Graphic Standards Planning Committee, State-of-the-Art Subcommittee," *Computer Graphics* 12, nos. 1–2 (June 1978): 14–169. For
a discussion of the variety of mapping functions for which conventional programs have been written, see Kurt Brassel, "A Survey of
Cartographic Display Software," *International Yearbook of Cartography* 17 (1977): 60–77.

48 ROMs and other chips containing software necessitate the term *firmware*, defined as hardware containing nonerasable software. The buyer
of firmware purchases permanent information, in contrast to the pur-
chaser of software for whom magnetic tape is merely a physical means
of transport for programs or data.

49 According to Turing's theorem, a fundamental principle in the theory
of automata, any process that can be decomposed onto a finite number of discrete steps can be carried out by an appropriately designed
digital computer. Cartographic design and mapmaking are both
computable processes. See Mark S. Monmonier, *Computer-assisted
Cartography: Principles and Prospects* (Englewood Cliffs, N.J.:
Prentice-Hall, 1982), pp. 29–31.

50 SYMAP, a computer program producing contour and choropleth maps
on the line printer, enjoyed enormous success in the late 1960s and
early 1970s. A grant from the Ford Foundation and several confer-
ences sponsored by Harvard University's Laboratory for Computer
Graphics and Spatial Analysis, SYMAP's publisher, made many potential users aware of the program's versatility. Over 600 institutions, mostly colleges and universities, installed SYMAP, which in
turn was responsible for much of the early interest in computer-assisted
cartography. See Timothy R. Petersen, *SYMAP: A Study of the Recent History of Cartography and Institutional Communication in
Computer-assisted Cartography*, unpublished masters thesis, Department of Geography, Syracuse University, 1982. A more recent,
most impressive Harvard software product is the ODYSSEY system.
See Eric Teicholz, "Geographic Information Systems: the ODYS-
SEY Project," *Journal of the Surveying and Mapping Division,
American Society of Civil Engineers* no. 106 (November 1980):
119–35.

51 SYMAP was notorious for its defaults, which led naive users to produce without much thought thousands of contour maps with the range
between the lowest and highest data values divided into five equal-
interval categories. It would have been far more appropriate had users
been encouraged to examine a frequency historgram or use a numerical procedure to search for natural breaks or, better yet, to invoke

all inherently meaningful breaks and use round-numbers for the remaining breaks. See Elri Liebenberg, "Symap: Its Uses and Abuses," *Cartographic Journal* 13 (1976): 26–36.

52 See Monmonier, *Computer-assisted Cartography*, pp. 154–77.

53 For a well-illustrated description of SAS/GRAPH see Kathryn A. Council and Jane T. Helwig, eds., *SAS/GRAPH User's Guide* (Cary, N.C.: SAS Institute, 1981).

CHAPTER 6: Map Publishing and the Digital Map

1 R. A. Skelton, *Maps: A Historical Survey of Their Study and Collecting* (Chicago: University of Chicago Press, 1972), pp. 12–13.

2 James Burke, *Connections* (Boston: Little, Brown and Co., 1978), pp. 101–4.

3 Arthur H. Robinson, "Mapmaking and Map Printing: The Evolution of a Working Relationship." In David Woodward, ed., *Five Centuries of Map Printing* (Chicago: University of Chicago Press, 1975), pp. 1–23 [ref., p. 7].

4 See Phil Davis, *Photography* (Dubuque, Iowa: William C. Brown, 1972), pp. 4–5.

5 See, for example, William C. Maxwell, *Printmaking: A Beginning Handbook* (Englewood Cliffs, N.J.: Prentice-Hall, 1977), pp. 203–71.

6 Walter W. Ristow, "Lithography and Maps, 1796–1850." In David Woodward, ed., *Five Centuries of Map Printing* (Chicago: University of Chicago Press, 1975), pp. 77–112.

7 C. Koeman, "The Application of Photography to Map Printing and the Transition to Offset Lithography." In David Woodward, ed., *Five Centuries of Map Printing* (Chicago: University of Chicago Press, 1975), pp. 137–55.

8 See, for example, James Craig, *Production for the Graphic Designer* (New York: Watson-Guptill Publications, 1974), pp. 86–89.

9 J. S. Keates, *Cartographic Design and Production* (London: John Wiley and Sons, 1973), pp. 127–30.

10 Robinson, "Mapmaking and Map Printing: The Evolution of a Working Relationship," pp. 7–8, 14.

11 See David Woodward, *The All-American Map: Wax Engraving and Its Influence on Cartography* (Chicago: University of Chicago Press, 1977).

12 Robinson, "Mapmaking and Map Printing: The Evolution of a Working Relationship," p. 22.

13 Ibid., p. 4.
14 Karen S. Pearson, "The Nineteenth-Century Colour Revolution: Maps in Geographical Journals," *Imago Mundi* 32 (1980): 9–20.
15 The coated side of the scribecoat faces upward if a right-reading image on the scribed sheet is to yield a right-reading image on the exposed emulsion. This is called "shooting through the base." A normal contact exposure, coated side to emulsion, will reverse the orientation of the image—from right-reading to wrong-reading, or from wrong-reading to right-reading.
16 See Koeman, "The Application of Photography to Map Printing and the Transition to Offset Lithography," pp. 153–54. The adoption of scribing and negative artwork in general has been rapid. Not mentioned in the first, 1953 edition of Arthur Robinson's widely used textbook *Elements of Cartography,* scribing received a three-page treatment in the second, 1960 edition. See Arthur H. Robinson, *Elements of Cartography,* 1st ed. (New York: John Wiley and Sons, 1953); and *Elements of Cartography,* 2nd ed. (New York: John Wiley and Sons, 1960), pp. 274–76. For a fuller discussion of scribing and its use in preparing topographic maps, see Lionel C. Moore, *Cartographic Scribing Materials, Instruments and Techniques* (Washington: American Congress on Surveying and Mapping, Technical Monograph no. CA–3, 1975).
17 Burke, *Connections,* p. 212.
18 Ibid., pp. 108–13.
19 See, for example, Mark S. Monmonier, *Computer-assisted Cartography: Principles and Prospects* (Englewood Cliffs, N.J.: Prentice-Hall, 1982), pp. 25–26.
20 Richard E. Matick, *Computer Storage Systems and Technology* (New York: John Wiley and Sons, 1977), pp. 14–15.
21 See, for example, S. Middlehoek, P. K. George, and P. Dekker, *Physics of Computer Memory Devices* (New York: Academic Press, 1976), pp. 7–11, 367–75.
22 For a review of direct access magnetic storage systems, see Matick, *Computer Storage Systems and Technology,* pp. 342–439; and Robert M. White, "Disk-Storage Technology," *Scientific American* 243, no. 2 (August 1980): 138–48.
23 See Hsu Chang, *Magnetic-Bubble Memory Technology* (New York: Marcel Dekker, 1978).
24 As used in discussions of computer memory, K represents not 1,000, but 1,024—the 10th power of 2. A word is a group of bits that are stored and retrieved as a unit; all bits allocated to a particular num-

ber, for example, the daily temperature, are thus treated as a unit. For a comprehensive discussion of problems encountered in manufacturing semiconductor memories, see Arthur L. Robinson, "Problems with Ultraminiaturized Transistors," *Science* 208 (1980): 1246–49; and John Walsh, "Japan-U.S. Competition: Semiconductors Are the Key," *Science* 215 (1982): 825–29.

25 Jerry Gray, "Implications of the Shuttle: Our Business in Space," *Technology Review* 84, no. 1 (October 1981): 34–46.

26 Lewis M. Branscomb, "Information: The Ultimate Frontier," *Science* 203 (1979): 143–47.

27 For a general discussion of videodisc technology, see Yuri Gates, "A Note on Videodiscs." In Philip Hills, ed., *The Future of the Printed Word: The Impact and Implications of the New Communications Technology* (Westport, Conn.: Greenwood Press, 1980), pp. 145–48; and Charles M. Goldstein, "Optical Disk Technology and Information," *Science* 215 (1982): 862–68.

28 Gates, "A Note on Videodiscs," p. 148.

29 For a discussion of the relative merits for cartographic applications of videodiscs and other storage media, see Richard K. Burkard, "Data Storage Technology Assessment," *Technical Papers of the American Congress on Surveying and Mapping* Fall Technical Meeting, 1980, paper no. CD–2–D, 9 pp. For a discussion of the use of computers to access selectively any of the 54,000 frames on a videodisc file, see Stuart Silverstone, "Interactive Videodisc Technology Applications," *Computer Graphics World* 6, no. 12 (December 1983): 59–62.

30 These figures, which represent a composite of the capabilities of the modern pen plotter, are derived from Mike Higgins, "1982 Graphic Systems Review," *Computer Graphics World* 5, no. 2 (February 1982): 65–72.

31 This paragraph and the one following describe the laser scanner-plotter manufactured by Scitex, an Israeli firm. Scitex plotters are used by large commercial printers for generating final negatives of both text and artwork, including such diverse applications as wallpaper and maps. See, for example, "On-line with Sci-Tex," *Graphic Arts Buyer*, 12, no. 5 (September–October 1981): 46–48; and Doyle G. Smith, "Raster Data Development in the National Mapping Division, U.S. Geological Survey," *Technical Papers of the American Congress on Surveying and Mapping* Fall Technical Meeting, 1981, pp. 284–88. For a discussion of plotting computer-generated screens on color separated final negatives, see P. Stefanovic, "Digital Screening

Techniques," *ITC Journal* no. 1982–2 (1982): 139–44. One cartographer believes 100 micrometer resolution, with 100 dots per cm
[250 dots per in.] is the coarsest acceptable screen tint for high-
quality cartographic reproduction; see Koert Sijmons, "Cartographic
Applications of Digital Techniques," *ITC Journal* no. 1982–2 (1982):
131–38. For a description of a commercial application of a scan dig-
itizer-plotter, see Martin Bruner, "The Production and Update of Road
Maps by Means of Computer-Assisted Procedures," *International
Yearbook of Cartography* 21 (1981): 23–29; and Michael L. Sena,
"Computer Mapping for Publication," *Computer Graphics World* 6,
no. 7 (July 1983): 68–76.

32 For a description of drum scanners and their operation, see John S.
    Montouri, "Image Scanner Technology," *Photogrammetric Engi-
    neering and Remote Sensing* 46 (1980): 49–61.

33 For a review of algorithms to convert raster data to a vector format,
    see Donna J. Peuquet, "An Examination of Techniques for Refor-
    matting Digital Cartographic Data. Part 1: The Raster-to-Vector
    Process," *Cartographica* 18, no. 1 (Spring 1981): 34–48.

34 For a description of a typical vector-mode digitizer and its use, see
    K. J. Dueker and R. H. Ericksen, "Interactive Digitizing and Graphics:
    University of Iowa Geography Graphics Laboratory," *Geo-Processing*
    1 (1979): 71–83. An automatic voice decoder used to control a dig-
    itizer is described in Bruno Beek and others, "Voice Data Entry for
    Cartographic Applications," *Proceedings of the American Congress
    on Surveying and Mapping* Fall Technical Meeting, 1977, pp.
    161–85.

35 Times presented here as examples are based on the author's experi-
    ence and discussions with representatives of manufacturers and per-
    sonnel in government mapping agencies. For a discussion of the rel-
    ative merits and deficiencies of raster-scan and line-following
    digitization, see A. R. Boyle, "Development in Equipment and
    Techniques." In D. R. Fraser Taylor, ed., *The Computer in Contem-
    porary Cartography* (Chichester: John Wiley and Sons, 1980), pp.
    39–57. An official of a prominent British mapmaking firm encour-
    ages the use of service bureaus for computer processing of digital
    cartographic data for maps produced by small map publishers; see
    Alan Poynter, "Computer-assisted Cartography—The Dilemma Fac-
    ing British Producers of Derived Mapping in the 1980s," *Carto-
    graphic Journal* 20 (1983): 31–34.

36 For a review of algorithms for converting vector data to raster for-
    mat, see Donna J. Peuquet, "An Examination of Techniques for Re-

formatting Digital Cartographic Data. Part 2: The Vector-to-Raster Process," *Cartographica* 18, no. 3 (Autumn 1981): 21–33.

37 Higgins, "1982 Graphics Systems Review," p. 71. Lines can vary in length from about 30 cm [1 ft] to over a meter. For a description of an electrostatic plotter able to produce color hardcopy 17 cm [42 in.] wide with a resolution of 80 dots per cm [200 dots per in.], see Mike Higgins, "Versatec's Color Electrostatic Plotter," *Computer Graphics World* 6, no. 7 (July 1983): 63–64.

38 For a discussion of this and other electrostatic printing methods, see Donald S. Swatik, "Nonimpact Printing." In A. D. Moore, ed., *Electrostatics and Its Applications* (New York: John Wiley and Sons, 1973), pp. 307–35.

39 See C. H. Hertz and T. Orhaug, "The Ink Jet Plotter: A Computer Peripheral for Producing Hard Copy Color Imagery," *Computer Graphics and Image Processing* 5 (1976): 1–12; and Larry Kuhn and Robert A. Myers, "Ink-Jet Printing," *Scientific American* 240, no. 4 (April 1979): 162–78.

40 See Wesley R. Iversen, "Laser Printers Head for the Office," *Electronics* 53, no. 2 (27 January 1981): 100–101; and J. Robert Lineback, "Net Gets Color Graphics Laser Printer," *Electronics* 54, no. 23 (17 November 1981): 154–56.

41 See, for example, James B. Brinton, "Dot-matrix Printer 'Hues' the Line," *Electronics* 54, no. 23 (17 November 1981): 47–48. For a comparison of a number of color graphics hardcopy devices, see Catherine Cramer, "Color Graphics Hard Copy Comes of Age," *Computer Graphics World* 6, no. 1 (January 1983): 29–34; and Mike Higgins, "Surveying the Color Hard-Copy Industry," *Computer Graphics World* 6, no. 1 (January 1983): 37–42.

42 See Joseph L. Kish, Jr., *Micrographics: A User's Manual* (New York: John Wiley and Sons, 1980), pp. 140–66.

43 For examples of cartographic applications of COM recorders, see T. A. Adams, H. M. Mounsey, and D. W. Rhind, "Topographic Maps from Computer Output on Microfilm," *Cartographic Journal* 17 (1980): 33–39; and Morton A. Meyer, Frederick R. Broome, and Richard H. Schweitzer, Jr., "Color Statistical Mapping by the U.S. Bureau of the Census," *American Cartographer* 2 (1975): 100–117.

44 In a series of matrices developed to classify existing and potential research themes in the history of cartography, Woodward recognizes document distribution as one of four principal stages of cartographic communication. Most of the comparatively rare studies of map distribution deal with the map trade. See David Woodward, "The Study

of the History of Cartography: A Suggested Framework," *American Cartographer* 1 (1974): 101–15.

45 Toffler sees the "de-massifying" of the mass media as an inevitable consequence of improved telecommunications. Through expanded and ungraded cable systems, television viewers will have access to a much wider variety of programs, some national and many local, but with most catering to special interest groups. See Alvin Toffler, *The Third Wave* (New York: William Morrow and Co., 1980), pp. 171–83.

46 James Martin, *Future Developments in Telecommunications,* 2nd ed. (Englewood Cliffs, N.J.: Prentice-Hall, 1977). pp. 60–65.

47 See Martin, *Future Developments in Telecommunications*, pp. 454–70; C. P. Sandback, ed., *Optical Fibre Communication Systems* (Chichester: John Wiley and Sons, 1980); W. S. Boyle, "Light-wave Communications," *Scientific American* 237, no. 2 (August 1977): 40–48; and Richard A. Cerny and Tad Witkowicz, "Fiber Optics in CAD/CAM Systems," *Computer Graphics World* 6, no. 5 (May 1983): 74–78.

48 See James Martin, *Telematic Society: A Challenge for Tomorrow* (Englewood Cliffs, N.J.: Prentice-Hall, 1981), pp. 110–20.

49 See Murray Laver, *Computers, Communications and Society* (London: Oxford University Press, 1975), pp. 42–46; and Martin, *Future Developments in Telecommunications,* pp. 539–56.

50 For a general introduction to videotex, see Kathleen Criner and Martha Johnson-Hall, "Videotex: Threat or Opportunity," *Special Libraries* 71 (1980): 379–85; and Jan Gecsei, *The Architecture of Videotex Systems* (Englewood Cliffs, N.J.: Prentice-Hall, 1983).

51 See "Teletext: TV Gets Married to the Printed Word," *Broadcasting* 97, no. 8 (20 August 1979): 30–36; Christopher P. Dingman, "Flick a Switch, Read the Newspaper," *Advertising Age* 50, no. 49, sec. 2 (19 November 1979): 36–38; and William J. Broad, "Upstart Television: Postponing a Threat," *Science* 210 (1980): 611–15.

52 See Susan Spaeth Cherry, "Telereference: The New TV Information Systems," *American Libraries* 11, no. 2 (February 1980): 94–98, 108–10; John Martyn, "Prestel and Public Libraries: An LA/Aslib Experiment," *ASLIB Proceedings* 31, no. 5 (May 1979): 216–36; and Andrew Pollack, "Phone-Cable 'Hybrids' Seen," *New York Times* 5 November 1981, p. D2. For an example of the digital transmission and processing of newspaper photographs, see "The Digital Photograph," *Editor and Publisher* 116, no. 47 (19 November 1983): 24–25, 27.

53 For a concise review of salient points in the federal regulation of videotex, see Richard M. Neustadt, Gregg P. Skall, and Michael Hammer, "The Regulation of Electronic Publishing," *Federal Communications Law Journal* 33 (1981): 331–417. For discussion of the problem of graphics standards for videotex, see G. Berton Latamore, "Graphics in Videotex," *Computer Graphics World* 6, no. 9 (September 1983): 58–62, 98.

54 Criner and Johnson-Hall, "Videotex: Threat or Opportunity," p. 382.

55 Neustadt, Skall, and Hammer, "The Regulation of Electronic Publishing," p. 400; and D. R. F. Taylor, "The Cartographic Potential of Telidon," *Cartographica* 19, nos. 3/4 (Autumn/Winter 1982): 18–30.

56 Martin, *Telematic Society: A Challenge for Tomorrow,* pp. 124–27.

57 For a discussion of the effects of electonics on publishers in general, see Robyn Shotwell, "Networking Is the Promise—and Problem—of Computers," *Publishers Weekly* 220, no. 17 (23 October 1981): 19–28.

58 See Octave Uzanne, "The End of Books," *Scribner's Magazine* 16, no. 2 (August 1894): 224–31 [reprinted in *Printing History* 1, no. 2 (1979): 23–32]. An amusing but insightful essay by R. J. Heathorn, titled "Learn with BOOK," describes the convenience and efficiency of printed books. In Philip Hills, ed., *The Future of the Printed Word: The Impact and the Implications of the New Communications Technology* (Westport, Conn.: Greenwood Press, 1980), pp. 171–72.

59 See, for example, Frederic Golden, Philip Faflick, and J. Madeline Nash, "Here Come the Microkids," *Time* 119, no. 18 (3 May 1982): 50–56.

60 Robinson, "Mapmaking and Map Printing: The Evolution of a Working Relationship," p. 4.

CHAPTER 7: Summary and Conclusions

1 Commissioners for the Internal Improvement of the State, *Report of the Commissioners Appointed by the Legislature of New York on the 8th of March, 1814, for the Internal Improvement of the State,* Readex Microprint Early American Imprint series, microcard 32,325, p. 3.

2 "Dashboard Navigator," *Science Digest* 90, no. 7 (July 1982): 24.

3 For a discussion of the problems of developing large integrated software systems, see Frederick P. Brooks, Jr., *The Mythical Man-Month: Essays on Software Engineering* (Reading, Mass.: Addison-Wesley Publishing Co., 1975).

4 For a discussion of self-publishing as promoted by videotex, see Ralph Lee Smith, "Processed Words: Plugging In to the New Technology," *Nation* 233 (1981): 313–15.

5 See, for examples, Anthony Smith, *The Politics of Information: Problems of Policy in Modern Media* (London: Macmillan, 1978), pp. 26–27; John Wicklein, *Electronic Nightmare: The New Communications and Freedom* (New York: Viking Press, 1981), pp. 100–155; and Willis H. Ware, *Public Policy Aspects for an Information Age* (Santa Monica, Calif.: The Rand Corporation, Rand Papers Series no. 5784, 1977).

6 See Michael Carley, *Social Measurement and Social Indicators: Issues of Policy and Theory* (London: George Allen and Unwin, 1981), pp. 12–13 and 173–74.

7 See James L. Cerny, "Awareness of Maps as Objects for Copyright," *American Cartographer* 5 (1978): 45–56.

8 Barry D. Rein, "Protecting the Content of a ROM—The Current State of the Law," *Computer Graphics News* 1, no. 4 (January/February 1982): 8.

9 Nicholas Henry, *Copyright—Information Technology—Public Policy, Part I: Copyright—Public Policies* (New York: Marcel Dekker, 1975), p. 75. For a discussion of British copyright practices and digital maps, see John Davies, "Copyright and the Electronic Map," *Cartographic Journal* 19 (1982): 135–36.

10 See M. Mitchell Waldrop, "Imaging the Earth (I): The Troubled First Decade of Landsat," *Science* 215 (1982): 1600–1603.

11 Council of State Governments, *Environmental Resource Data: Intergovernment Management Dimensions* (Lexington, Ky.: Council of State Governments, 1978), p. 10.

12 R. B. Southard, "The Development of U.S. National Mapping Policy," *American Cartographer* 10 (1983): 5–15. A single federal mapping agency most likely would have avoided the significant waste from two or more agencies independently producing similar digital cartographic data bases, as noted in a recent report by the U.S. General Accounting Office; see Comptroller General of the United States, *Duplicative Federal Computer-Mapping Programs: A Growing Problem* (Washington, D.C.: General Accounting Office, report no. GAO/RCED–83–19, November 22, 1982).

13 Panel to Review the Report of the Federal Mapping Task Force on Mapping, Charting, Geodesy and Surveying, July 1973, Committee on Geodesy, Assembly of Mathematical and Physical Sciences, National Research Council, *Federal Surveying and Mapping: An Or-*

ganizational Review (Washington: National Academy Press, 1981), pp. 3–5.

14 Edward Higbee, *A Question of Priorities: New Strategies for Our Urbanized World* (New York: William Morrow and Co., 1970), pp. 80–81.

15 See Richard Goody, "Satellites for Oceanography—The Promises and the Realities," *Oceanus* 24, no. 3 (Fall 1981): 2–5. The entire Fall 1981 issue of *Oceanus* is devoted to oceanography and remote sensing applications.

16 William J. Broad, "Nuclear Pulse (I): Awakening to the Chaos Factor," *Science* 212 (1981): 1009–12.

17 Gerald Steinberg, "The Ultimate Battleground: Weapons in Space," *Technology Review* 84, no. 1 (October 1981): 57–63.

18 Ed Zuckerman, "How Would the U.S. Survive a Nuclear War?" *Esquire* 97, no. 3 (March 1982): 37–46.

19 Harold Weiss, "Computer Security: An Overview," *Datamation* 20, no. 1 (January 1974): 42–47.

20 S. M. Miranda, "Aspects of Data Security in General-purpose Data Base Management Systems." In *Proceedings of the 1980 Symposium on Security and Privacy, April 14–16, 1980, Oakland, California* (New York: Institute of Electrical and Electronic Engineers, 1980), pp. 46–58; and Ivars Peterson, "Computer Crime: Insecurity in Numbers," *Science News* 122 (1982): 12–14.

21 John M. Carroll, *Data Base and Computer Systems Security* (Wellesley, Mass.: Q. E. D. Information Sciences, 1976), pp. 42–45.

22 Richard R. Linde, "Operating System Penetration," *Proceedings of the 1975 National Computer Conference* (Montvale, N.J.: AFIPS Press, 1975), pp. 361–68.

23 P. Tucker Withington, "The Trusted Function in Secure Decentralized Processing." In *Proceedings of the 1980 Symposium on Security and Privacy, April 14–16, 1980, Oakland, California* (New York: Institute of Electrical and Electronic Engineers, 1980), pp. 67–79.

24 Charles F. Hemphill, Jr., and Robert D. Hemphill, *Security Safeguards for the Computer* (New York: AMACOM, a division of American Management Associations, 1979), pp. 25–30.

25 See American Federation of Information Processing Societies, *American Federation of Information Processing Societies System Review Manual on Security* (Montvale, N.J.: American Federation of Information Processing Societies, 1974); Lance J. Hoffman, *Modern Methods for Computer Security and Privacy* (Englewood Cliffs, N.J.:

Prentice-Hall, 1977); and David K. Hsiao, Douglas S. Kerr, and Stuart E. Madnick, *Computer Security* (New York: Academic Press, 1979).

26 Francis W. Dolloff and Roy L. Perkinson, *How to Care for Works of Art on Paper,* 3rd ed. (Boston: Museum of Fine Arts, 1979), pp. 7–24.

27 See Lee E. Grove, "Paper Deterioration—An Old Story," *College and Research Libraries* 25 (1964): 365–74.

28 See Mary Larsgaard, *Map Librarianship: An Introduction* (Littleton, Colo.: Libraries Unlimited, 1978), pp. 156–58; and Richard Daniel Smith, "Maps, Their Deterioration and Preservation," *Special Libraries* 63 (1972): 59–68.

29 William K. Wilson and others, "The Effect of Magnesium Bicarbonate Solutions on Various Papers." In John C. Williams, ed., *Preservation of Papers and Textiles of Historic and Artistic Value II* (Washington: American Chemical Society, 1981), pp. 87–107.

30 Peter Waters, "Archival Methods of Treatment for Library Documents." In John C. Williams, ed., *Preservation of Papers and Textiles of Historic and Artistic Value II* (Washington: American Chemical Society, 1981), pp. 13–23.

31 P. S. Davison, P. Giles, and D. A. R. Matthews, "Aging of Magnetic Tape: A Critical Bibliography and Comparison of Literature Sources," *Computer Journal* 11 (1968): 241–46.

32 Lewis M. Branscomb, "Information: The Ultimate Frontier," *Science* 203 (1979): 143–47.

33 Joseph P. Martino, "Telecommunications in the Year 2000," *Futurist* 13, no. 2 (April 1979): 95–103.

34 Ted Becker, "Teledemocracy: Bringing Power Back to People," *Futurist* 15, no. 6 (December 1981): 6–9.

35 Hollis Vail, "The Home Computer Terminal: Transforming the Household of Tomorrow," *Futurist* 14, no. 6 (December 1980): 52–58.

36 Henry B. Freedman, "Paper's Role in an Electronic World," *Futurist* 15, no. 5 (October 1981): 11–16.

37 Alvin Toffler, *The Third Wave* (New York: William Morrow and Co., 1980), pp. 210–23.

38 Robert D. Hamrin, "The Information Economy: Exploiting an Infinite Resource," *Futurist* 15, no. 4 (August 1981): 25–30.

39 Mark S. Monmonier, "Map-Text Coordination in Geographic Writing," *Professional Geographer* 33 (1981): 406–12.

40 Mark S. Monmonier, "Cartography, Geographic Information, and

Public Policy," *Journal of Geography in Higher Education* 6 (1982): 99–107.

41 Ronald R. Mourant, Raman Lakshmanan, and Roongrojn Chantadisai, "Visual Fatigue and Cathode-ray Tube Display Terminals," *Human Factors* 23 (1981): 529–40.

42 See "Face Rashes Linked with Use of VDTs," *Science News* 120 (1981): 150; I. B. McKee, "Health and Psychological Aspects Related to the Operation of Computer Workstations." In David Rhind and Tim Adams, eds., *Computers in Cartography* (London: British Cartographic Society, Special Publication no. 2, 1982), pp. 43–50; Eliot Marshall, "FDA Sees No Radiation Risk in VDT Screens," *Science* 212 (1981): 1120–21; "Safety Rules Recognize New Hazard in Screens," *New Scientist* 91 (1981): 393; and "VDTs and the Fetus—Cause for Concern?" *Science News* 120 (1981): 377. A survey by the Newspaper Guild suggests that VDT users suffer from higher than average rates of neck pain, shoulder pain, and lower back pain; see James E. Roper, "VDTs and Employee Productivity," *Editor and Publisher* 116, no. 43 (22 October 1983): 34.

43 See David Woodward, "Introduction" to *Cartography and Art* (Chicago: University of Chicago Press, forthcoming).

44 See John Lewell, "Computers Extend the Artist's Horizon," *New Scientist* 92 (1981): 750–54; and Rodney Stock, "Introduction to Digital Computer Graphics," *SMPTE Journal* 90 (1981): 1184–89.

Bibliography

Adams, T. A., Mounsey, H. M., and Rhind, D. W. "Topographic Maps from Computer Output on Microfilm." *Cartographic Journal* 17 (1980): 33–39.

Alterman, Hyman. *Counting People: The Census in History.* New York: Harcourt, Brace and World, 1969.

American Federation of Information Processing Societies. *AFIPS System Review Manual on Security.* Montvale, N.J.: American Federation of Information Processing Societies, 1974.

Anderson, James R. "The National Atlas of the United States." In Barbara J. Gutsell, ed. *The Purpose and Use of National Atlases.* [Also *Cartographica,* monograph no. 23.] Toronto: University of Toronto Press, 1979, pp. 35–39.

Anderson, James R., and others. *A Land Use and Land Cover Classification System for Use with Remote Sensor Data.* Washington: U.S. Government Printing Office, U.S. Geological Survey Professional Paper no. 964, 1976.

Andrews, J. H. *A Paper Landscape: The Ordnance Survey in Nineteenth-century Ireland.* Oxford: Oxford University Press, 1975.

Babington-Smith, Constance. *Air Spy.* New York: Harper and Row, 1957.

Baker, L. Ralph, and others. "Electro-Optical Remote Sensors with Related Optical Sensors." In *Manual of Remote Sensing.* Falls Church, Va.: American Society of Photogrammetry, 1975, pp. 326–66.

Bannister, A., and Raymond, S. *Surveying,* 4th ed. London: Pitman Publishing, 1977.

Barrett, Eric C., and Hamilton, Michael G. "The Use of Geostationary Satellite Data in Environmental Science." *Progress in Physical Geography* 6 (1982): 159–214.

Bean, Russell K. "The Orthophotoscope and Its Development." *Canadian Surveyor* 22 (1968): 38–45.

Becker, Ted. "Teledemocracy: Bringing Power Back to People." *Futurist* 15, no. 6 (December 1981): 6–9.

229

Beek, Bruno, and others. "Voice Data Entry for Cartographic Applications." *Proceedings of the American Congress on Surveying and Mapping,* Fall Technical Meeting, 1977, pp. 161–85.

Bell, Timothy P., "A Practical Approach to Electronic Distance Measurement." *Surveying and Mapping* 38 (1978): 335–41.

Benedict, George H. "Map Engraving." *Printing Art* 19 (1912): 205–8.

Benny, A. H., and Dawson, G. J. "Satellite Imagery as an Aid to Bathymetric Charting in the Red Sea." *Cartographic Journal* 20 (1983): 5–16.

Bernstein, George B., and Kashar, Arnold S. *Intelligent Terminals: Functions, Specifications, and Applications.* Wellesley, Mass.: Q. E. D. Information Sciences, 1978.

Blakemore, Michael. "From Way-finding to Map-making: The Spatial Information Fields of Aboriginal Peoples." *Progress in Human Geography* 5 (1981): 1–24.

Blakemore, Michael, and Harley, J. B. *Concepts in the History of Cartography: A Review and Perspective.* [Also *Cartographica,* monograph no. 26.] Toronto: University of Toronto Press, 1980.

Bossler, John D. "A Note on Global Positioning System Activities." *Bulletin of the American Congress on Surveying and Mapping* no. 74 (1981): 39–40.

Bossler, John D. "New Adjustment of North American Datum." *Journal of the Surveying and Mapping Division, Proceedings of the American Society of Civil Engineers* 108, no. SU2 (August 1982): 47–52.

Bowditch, Nathaniel. *American Practical Navigator: An Epitome of Navigation,* corrected print. Washington: U.S. Government Printing Office, 1966.

Boyle, A. R. "Development in Equipment and Techniques." In D. R. Fraser Taylor, ed. *The Computer in Contemporary Cartography.* Chichester: John Wiley and Sons, 1980, pp. 39–57.

Branscomb, Lewis M. "Information: The Ultimate Frontier." *Science* 203 (1979): 143–47.

Brassel, Kurt. "A Survey of Cartographic Display Software." *International Yearbook of Cartography* 17 (1977): 60–77.

Brinker, Russell C., and Wolf, Paul R. *Elementary Surveying,* 6th ed. New York: IEP—A Dun-Donnelley Publisher, 1977.

Brinton, James B. "Dot-matrix Printer 'Hues' the Line." *Electronics* 54, no. 23 (17 November 1981): 47–48.

Broad, William J. "Nuclear Pulse (I): Awakening to the Chaos Factor." *Science* 212 (1981): 1009–12.

Broad, William J. "Upstart Television: Postponing a Threat." *Science* 210 (1980): 611–15.

Brooks, Frederick P., Jr. *The Mythical Man-Month: Essays on Software Engineering*. Reading, Mass.: Addison-Wesley Publishing Co., 1975.

Brown, Bruce Eric. "Computer Graphics for Large Scale Two- and Three-Dimensional Analysis of Complex Geometries." *Computer Graphics* 13, no. 2 (August 1979): 31–40.

Brown, Lloyd A. *The Story of Maps*. Boston: Little, Brown and Co., 1949.

Bruner, Martin. "The Production and Update of Road Maps by Means of Computer-Assisted Procedures." *International Yearbook of Cartography* 21 (1981): 23–29.

Burkard, Richard K. "Data Storage Technology Assessment." *Technical Papers of the American Congress on Surveying and Mapping*, Fall Technical Meeting, 1980, paper no. CD–2–D.

Burke, James. *Connections*. Boston: Little, Brown and Co., 1978.

Cardwell, D. S. L. *Turning Points in Western Technology*. New York: Science History Publications, 1972.

Carley, Michael. *Social Measurement and Social Indicators: Issues of Policy and Theory*. London: George Allen and Unwin, 1981.

Carroll, John M. *Data Base and Computer Systems Security*. Wellesley, Mass.: Q. E. D. Information Sciences, 1976.

Castner, Henry W. "Concept Before Content? A Question in Atlas Design with Special Reference to the National Atlas of Canada." *Canadian Geographer* 20 (1976): 224–32.

Cerny, James L. "Awareness of Maps as Objects for Copyright." *American Cartographer* 5 (1978): 45–56.

Cerny, Richard A., and Witkowicz, Tad. "Fiber Optics in CAD/CAM Systems." *Computer Graphics World* 6, no. 5 (May 1983): 74–78.

Chang, Hsu. *Magnetic-Bubble Memory Technology*. New York: Marcel Dekker, 1978.

Chapman, William H. "Proposed Specifications for Inertial Surveying." *Technical Papers of the American Congress on Surveying and Mapping*, 43rd Annual Meeting, 1983, pp. 287–93.

Cherry, Susan Spaeth. "Telereference: The New TV Information Systems." *American Libraries* 11, no. 2 (February 1980): 94–98, 108–10.

Chrzanowski, Adam, and others. "A Forecast of the Impact of GPS on Surveying." *Technical Papers of the American Congress on Surveying and Mapping*, 43rd Annual Meeting, 1983, pp. 625–34.

Clark, David. *Plane and Geodetic Surveying, Volume Two: Higher Surveying.* London: Constable, 1973.

Colcord, J. E. "The Surveying Engineer and NAD–83." *Journal of the Surveying and Mapping Division, American Society of Civil Engineers* 107, no. SU1 (November 1981): 25–31.

Colvocoresses, Alden P. "Applications to Cartography: Introduction." In *ERTS–1, A New Window on Our Planet.* Washington: U.S. Government Printing Office, U.S. Geological Survey Professional Paper No. 929, 1976, pp. 12–22.

Colvocoresses, Alden P. "The Relationship of Acquisition Systems to Automated Stereo Correlation." *Photogrammetric Engineering and Remote Sensing* 49 (1983): 539–44.

Colvocoresses, Alden P., and others. "Platforms for Remote Sensors." In *Manual of Remote Sensing.* Falls Church, Va.: American Society of Photogrammetry, 1975, pp. 539–88.

Combination Atlas of Bucks County, Pennsylvania. Philadelphia: J. D. Scott, 1876.

Combs, John E., and others. "Planning and Executing the Photogrammetric Project." In *Manual of Photogrammetry,* 4th ed. Falls Church, Va.: American Society of Photogrammetry, 1980, pp. 367–412.

Commission III—Computer-assisted Cartography, International Cartographic Association. *A Glossary of Technical Terms in Computer-assisted Cartography.* Falls Church, Va.: American Congress on Surveying and Mapping, for the International Cartographic Association, 1980.

Commissioners for the Internal Improvement of the State. *Report of the Commissioners Appointed by the Legislature of New York on the 8th of March, 1814, for the Internal Improvement of the State.* Readex Microprint Early American Imprint series, microcard 32,325.

Comptroller General of the United States. *Duplicative Federal Computer-Mapping Programs: A Growing Problem.* Washington, D.C.: General Accounting Office, report no. GAO/RCED–83–19, November 22, 1982.

"Cost-benefit Trips Up the Corps." *Business Week,* Industrial Edition no. 2573 (19 February 1979): 96–97.

Council, Kathryn A., and Helwig, Jane T., eds. *SAS/GRAPH User's Guide.* Cary, N.C.: SAS Institute, 1981.

Council of State Governments. *Environmental Resource Data: Intergovernmental Management Dimensions.* Lexington, Ky.: Council of State Governments, 1978.

Craig, James. *Production for the Graphic Designer.* New York: Watson-Guptill Publications, 1974.

Cramer, Catherine. "Color Graphics Hard Copy Comes of Age." *Computer Graphics World* 6, no. 1 (January 1983): 29–34.

Criner, Kathleen, and Johnson-Hall, Martha, "Videotex: Threat or Opportunity." *Special Libraries* 71 (1980): 379–85.

"Dashboard Navigator." *Science Digest* 90, no. 7 (July 1982): 24.

Davenport, William. "Marshall Islands Navigation Charts." *Imago Mundi* 15 (1960): 19–26.

Davies, John. "Copyright and the Electronic Map." *Cartographic Journal* 19 (1982): 135–36.

Davis, Phil. *Photography.* Dubuque, Iowa: William C. Brown, 1972.

Davis, Samuel. *Computer Data Displays.* Englewood Cliffs, N.J.: Prentice-Hall, 1969.

Davison, P. S., Giles, P., and Matthews, D. A. R. "Aging of Magnetic Tape: A Critical Bibliography and Comparison of Literature Sources." *Computer Journal* 11 (1968): 241–46.

"The Digital Photograph." *Editor and Publisher* 116, no. 47 (19 November 1983): 24–25, 27.

Dingman, Christopher P. "Flick a Switch, Read the Newspaper." *Advertising Age* 50, no. 49, sec. 2 (19 November 1979): 36–38.

Dobson, Jerome. "Automated Geography." *Professional Geographer* 35 (1983): 135–43.

Dolloff, Francis W., and Perkinson, Roy L. *How to Care for Works of Art on Paper,* 3rd ed. Boston: Museum of Fine Arts, 1979.

Doyle, Frederick J. "Digital Terrain Models: An Overview." *Photogrammetric Engineering and Remote Sensing* 44 (1978): 1481–84.

Doyle, Frederick J. "Satellite Systems for Cartography." *ITC Journal* no. 1981–2 (1981): 153–70.

Doyle, Frederick J. "The Next Decade of Satellite Remote Sensing." *Photogrammetric Engineering and Remote Sensing* 44 (1978): 155–64.

Dueker, K. J., and Ericksen, R. H. "Interactive Digitizing and Graphics: The University of Iowa Geography Graphics Laboratory." *Geo-Processing* 1 (1979): 71–83.

Dunlap, G. D., and Shufeldt, H. H. *Dutton's Navigation and Piloting,* 12th ed. Annapolis, Md.: U.S. Naval Institute, 1969.

Dutton, Geoffrey. "American Graph Fleeting: A Computer-Holographic Map Animation." In *Computer Mapping in Education, Research, and Medicine.* Cambridge, Mass.: Laboratory for Computer Graphics and Spatial Analysis, Harvard University, 1979, pp. 53–62.

Eastman, J. R., Nelson, W., and Shields, G. "Production Considerations in Isodensity Mapping." *Cartographica* 18, no. 1 (Spring 1981): 24–30.

Elachi, Charles. "Spaceborne Imaging Radar: Geologic and Oceanographic Applications." *Science* 209 (1980): 1073–82.

Elassal, Atef A. "Generalized Adjustment by Least Squares (GALS)." *Photogrammetric Engineering and Remote Sensing* 49 (1983): 201–6.

Ellis, Melvin Y., ed. *Coastal Mapping Handbook.* Washington: U.S. Government Printing Office, 1978.

"Equation of Time." *Encyclopedia Britannica,* 11th ed. (1910).

Ewald, R. H., and Fryer, R., eds. "Final Report of the Graphic Standards Planning Committee, State-of-the-Art Subcommittee." *Computer Graphics* 12, nos. 1–2 (June 1978): 14–169.

Eyton, J. Ronald. "Landsat Multitemporal Color Composites." *Photogrammetric Engineering and Remote Sensing* 49 (1983): 231–35.

"Face Rashes Linked with Use of VDTs." *Science News* 120 (1981): 150.

Fischer, Irene. "Is the Astrogeodetic Approach in Geodesy Obsolete?" *Surveying and Mapping* 34 (1974): 121–30.

Fischer, William A., and others. "History of Remote Sensing." In *Manual of Remote Sensing,* vol. I. Falls Church, Va.: American Society of Photogrammetry, 1975, pp. 27–50.

Freedman, Henry B. "Paper's Role in an Electronic World." *Futurist* 15, no. 5 (October 1981): 11–16.

Friedman, S. Jack, and others. "Automation of the Photogrammetric Process." In *Manual of Photogrammetry,* 4th ed. Falls Church, Va.: American Society of Photogrammetry, 1980, pp. 699–722.

Froome, K. D., and Essen, L. *The Velocity of Light and Radio Waves.* London and New York: Academic Press, 1969.

Fu, King Sun, and Yu, T. S. *Statistical Pattern Classification Using Contextual Information.* New York: Research Studies Press, John Wiley and Sons, 1980.

Gannett, Henry. "The Mapping of the United States." *Scottish Geographical Magazine* 8 (1892): 150–53.

Gannett, Henry. "The Mother Maps of the United States." *National Geographic Magazine* 4 (1892): 101–16.

Gates, Yuri. "A Note on Videodiscs." In Philip Hills, ed. *The Future of the Printed Word: The Impact and Implications of the New Communications Technology.* Westport, Conn.: Greenwood Press, 1980, pp. 145–48.

Gecsei, Jan. *The Architecture of Videotex Systems.* Englewood Cliffs, N.J.: Prentice-Hall, 1983.

Gignilliat, Robert L. "Cleaning Up, Matching and Merging Data Files." *Review of Public Data Use* 4, no. 2 (March 1976): 9–15.

Giloi, Wolfgang K. *Interactive Computer Graphics.* Englewood Cliffs, N.J.: Prentice-Hall, 1978.

Goddard, George W., and Copp, DeWitt S. *Overview: A Lifelong Adventure in Aerial Photography.* New York: Doubleday, 1969.

Golden, Frederic, Faflick, Philip, and Nash, J. Madeline. "Here Come the Microkids." *Time* 119, no. 18 (3 May 1982): 50–56.

Goldstein, Charles. "Optical Disk Technology and Information." *Science* 215 (1982): 862–68.

Goody, Richard. "Satellites for Oceanography—The Promises and the Realities." *Oceanus* 24, no. 3 (Fall 1981): 2–5.

Gordon, Don E. *Electronic Warfare.* New York: Pergamon Press, 1981.

Gray, David H. "The Preparation of Loran-C Lattices for Canadian Charts." *Canadian Surveyor* 34 (1980): 277–95.

Gray, Jerry. "Implications of the Shuttle: Our Business in Space." *Technology Review* 84, no. 1 (October 1981): 34–36.

Greene, John R. "Accuracy Evaluation in Electro-Optical Distance-Measuring Instruments." *Surveying and Mapping* 37 (1977): 247–56.

Groot, Richard. "Canada's National Atlas Program in the Computer Era." In Barbara J. Gutsell, ed. *The Purpose and Use of National Atlases.* [Also *Cartographica,* monograph no. 23.] Toronto: University of Toronto Press, 1979, pp. 41–52.

Grove, Lee E. "Paper Deterioration—An Old Story." *College and Research Libraries* 25 (1964): 365–74.

Gutersohn, Heinrich. "Atlas der Schweiz—ein Rueckblick." *Geographica Helvetica* 34, no. 4 (1979): 181–88.

Halbouty, Michel T. "Geologic Significance of Landsat Data for 15 Giant Oil and Gas Fields." *American Association of Petroleum Geologists* 64 (1980): 8–36.

Hamrin, Robert D. "The Information Economy: Exploiting an Infinite Resource." *Futurist* 15, no. 4 (August 1981): 25–30.

Heathorn, R. J. "Learn with BOOK." In Philip Hills, ed. *The Future of the Printed Word: The Impact and the Implications of the New Communications Technology.* Westport, Conn.: Greenwood Press, 1980. pp. 171–72.

Hemphill, Charles F., Jr., and Hemphill, Robert D. *Security Safeguards for the Computer.* New York: AMACOM, a division of American Management Associations, 1979.

Henry, Nicholas. *Copyright—Information Technology—Public Policy, Part I: Copyright—Public Policies*. New York: Marcel Dekker, 1975.

Herriot, Roger A. "The 1980 Census: Countdown for a Complete Count." *Monthly Labor Review* 102, no. 9 (September 1979): 3–13.

Hertz, C. H., and Orhaug, T. "The Ink Jet Plotter: A Computer Peripheral for Producing Hard Copy Color Imagery." *Computer Graphics and Image Processing* 5 (1976): 1–12.

Higbee, Edward. *A Question of Priorities: New Strategies for Our Urbanized World*. New York: William Morrow and Co., 1970.

Higgins, Mike. "1982 Graphics Systems Review." *Computer Graphics World* 5, no. 2 (February 1982): 65–72.

Higgins, Mike. "Surveying the Color Hard-Copy Industry." *Computer Graphics World* 6, no. 1 (January 1983): 37–42.

Higgins, Mike. "Versatec's Color Electrostatic Plotter." *Computer Graphics World* 6, no. 7 (July 1983): 63–64.

Hirvonen, R. *Adjustment by Least Squares in Geodesy and Photogrammetry*. New York: Frederick Ungar Publishing Co., 1971.

Hobbie, Dierk, and Faust, Hans W. "Z–2 ORTHOCOMP, the New High Performance Orthophoto Equipment from Zeiss." *Photogrammetric Engineering and Remote Sensing* 49 (1983): 635–40.

Hoffman, Lance J. *Modern Methods for Computer Security and Privacy*. Englewood Cliffs, N.J.: Prentice-Hall, 1977.

Hopkin, V. David, and Taylor, Robert M. *Human Factors in the Design and Evaluation of Aviation Maps*. Neuilly Sur Seine: NATO, Advisory Group for Aerospace Research and Development, no. AGARD–AG–225, 1979.

Hord, R. Michael. "Digital Enhancement of Landsat MSS Data for Mineral Exploration." In William L. Smith, ed. *Remote Sensing Applications for Mineral Exploration*. Stroudsburg, Pa.: Dowden, Hutchinson, and Ross, 1977, pp. 235–50.

Hothem, Larry D., Strange, William E., and White, Madeline. "Doppler Satellite Surveying System." *Journal of the Surveying and Mapping Division, American Society of Civil Engineers* 104 (1978): 79–91.

How 'Tis Done: A Thorough Ventilation of the Numerous Schemes Conducted by Wandering Canvassers Together with the Various Advertising Dodges for the Swindling of the Public. Syracuse: W. I. Pattison, 1890.

Howse, Derek. *Greenwich Time and the Discovery of the Longitude*. Oxford: Oxford University Press, 1980.

Hsiao, David K., Kerr, Douglas S., and Madnick, Stuart E. *Computer Security.* New York: Academic Press, 1979.

Iversen, Wesley R. "Laser Printers Head for the Office." *Electronics* 53, no. 2 (27 January 1981): 100–101.

Jensen, Homer, and others. "Side-looking Airborne Radar." *Scientific American* 237, no. 4 (October 1977): 84–95.

Jensen, John R. "Biophysical Remote Sensing." *Annals of the Association of American Geographers* 73 (1983): 111–32.

Jensen, John R. "Urban Change Detection Mapping Using Landsat Data," *American Cartographer* 8 (1981): 127–47.

Johnson, Hildegard Binder. *Order Upon the Land: The U.S. Rectangular Land Survey and the Upper Mississippi Country.* New York: Oxford University Press, 1976.

Kaneto, Toyohisa. "Evaluation of LANDSAT Image Registration Accuracy." *Photogrammetric Engineering and Remote Sensing* 42 (1976): 1285–99.

Kazan, B. "Materials Aspects of Display Devices." *Science* 208 (1980): 927–37.

Keates, J. S. *Cartographic Design and Production.* London: John Wiley and Sons, 1973.

Kish, Joseph L., Jr. *Micrographics: A User's Manual.* New York: John Wiley and Sons, 1980.

Kley, Vic. "Pointing Device Communication." *Computer Graphics World* 6, no. 11 (November 1983): 69–72.

Kock, Winston E. *Lasers and Holography,* 2nd ed., enlarged. New York: Dover Publications, 1981.

Koeman, C. "The Application of Photography to Map Printing and the Transition to Offset Lithography." In David Woodward, ed. *Five Centuries of Map Printing.* Chicago: University of Chicago Press, 1975, pp. 137–55.

Koestler, Arthur. *The Act of Creation.* New York: Macmillan, 1964.

Kolata, Gina Bari. "Geodesy: Dealing with an Enormous Computer Task." *Science* 200 (1978): 421–22, 466.

Konvitz, Josef V. "Redating and Rethinking the Cassini Geodetic Surveys of France, 1730–1750." *Cartographica* 19, no. 1 (Spring 1982):1–15.

Kuhn, Larry, and Myers, Robert A. "Ink-Jet Printing." *Scientific American* 240, no. 4 (April 1979): 162–78.

Laboratory for Computer Graphics and Spatial Analysis. *Lab-Log 1980.* Cambridge, Mass.: Laboratory for Computer Graphics and Spatial Analysis, Harvard University, 1980.

Larsgaard, Mary. *Map Librarianship: An Introduction.* Littleton, Colo.: Libraries Unlimited, 1978.

Latamore, G. Berton. "Graphics in Videotex." *Computer Graphics World* 6, no. 9 (September 1983): 58–62, 98.

Laurila, Simo. *Electronic Surveying and Mapping.* Columbus: Ohio State University Press, Institute of Geodesy, Photogrammetry and Cartography, Publication No. 11, 1960.

Laver, Murray. *Computers, Communications and Society.* London: Oxford University Press, 1975.

Lehman, Maxwell, and Burke, Thomas J. M. *Communication Technologies and Information Flow.* New York: Pergamon Press, 1981.

Lewell, John. "Computers Extend the Artist's Horizon." *New Scientist* 92 (1981): 750–54.

Liebenberg, Elri. "Symap: Its Uses and Abuses." *Cartographic Journal* 13 (1976): 26–36.

Lienhard, John H. "The Rate of Technological Improvement Before and After the 1830s." *Technology and Culture* 20 (1979): 515–30.

Lillesand, Thomas M. "Issues Surrounding the Commercialization of Civil Remote Sensing from Space." *Photogrammetric Engineering and Remote Sensing* 49 (1983): 495–504.

Lillesand, Thomas M., and Kiefer, Ralph W. *Remote Sensing and Image Interpretation.* New York: John Wiley and Sons, 1979.

Linde, Richard R. "Operating System Penetration." In *Proceedings of the 1975 National Computer Conference.* Montvale, N.J.: AFIPS Press, 1975, pp. 361–68.

Lineback, J. Robert. "Net Gets Color Graphics Laser Printer." *Electronics* 54, no. 23 (17 November 1981): 154–56.

Lippold, H. R., Jr. "Readjustment of the National Geodetic Vertical Datum." *Surveying and Mapping* 40 (1980): 155–64.

Lowe, Donald S., and others. "Imaging and Nonimaging Sensors." In *Manual of Remote Sensing.* Falls Church, Va.: American Society of Photogrammetry, 1975, pp. 367–97.

Lulla, Kamlesh. "The Landsat Satellites and Selected Aspects of Physical Geography." *Progress in Physical Geography* 7 (1983): 1–45.

McEntyre, John G. *Land Survey Systems.* New York: John Wiley and Sons, 1978.

McGrath, Gerald. "Re-defining the Role of Government in Surveys and Mapping/A View of Events in the United Kingdom." *Cartographica* 19, nos. 3/4 (Autumn/Winter 1982): 44–52.

McGrath, James J. "Contemporary Map Displays." In North Atlantic Treaty Organization, Advisory Group for Aerospace Research and

Development. *Guidance and Control Displays,* AGARD Conference Proceedings No. 96, February 1972, pp. 13–1 to 13–16.

McKee, I. B. "Health and Psychological Aspects Related to the Operation of Computer Workstations." In David Rhind and Tim Adams, eds. *Computers in Cartography.* London: British Cartographic Society, Special Publication no. 2, 1982, pp. 43–50.

Maling, D. H. *Coordinate Systems and Map Projections.* London: George Philip and Son Limited, 1973.

Manchester, Harland. *New Trail Blazers of Technology.* New York: Charles Scribner's Sons, 1978.

Manouher, Naraghi, Stromberg, William, and Daily, Mike. "Geometric Rectification of Radar Imagery Using Digital Elevation Models." *Photogrammetric Engineering and Remote Sensing* 49 (1983): 195–99.

A Manual of the Principal Instruments Used in American Engineering and Surveying, Manufactured by W. and L. E. Gurley, Troy, N.Y., U.S.A., 29th ed. Troy, N.Y.: W. and L. E. Gurley, 1891.

Marshall, Eliot. "FDA Sees No Radiation Risk in VDT Screens." *Science* 212 (1981): 1120–21.

Martin, James. *Future Developments in Telecommunications,* 2nd ed. Englewood Cliffs, N.J.: Prentice-Hall, 1977.

Martin, James. *Telematic Society: A Challenge for Tomorrow.* Englewood Cliffs, N.J.: Prentice-Hall, 1981.

Martino, Joseph P. "Telecommunications in the Year 2000." *Futurist* 13, no. 2 (April 1979): 95–103.

Martyn, John, "Prestel and Public Libraries: An LA/Aslib Experiment." *ASLIB Proceedings* 31, no. 5 (May 1979): 216–36.

Matick, Richard E. *Computer Storage Systems and Technology.* New York: John Wiley and Sons, 1977.

Maxwell, William C. *Printmaking: A Beginning Handbook.* Englewood Cliffs, N.J.: Prentice-Hall, 1977.

Meyer, Morton A., Broome, Frederick R., and Schweitzer, Richard H., Jr. "Color Statistical Mapping by the U.S. Bureau of the Census." *American Cartographer* 2 (1975): 100–117.

Mezera, David F. "Trilateration Adjustment Using Unit Corrections Derived from Least Squares." *Surveying and Mapping* 43 (1983): 315–29.

Middelhoek, S., George, P. K., and Dekker, P. *Physics of Computer Memory Devices.* New York: Academic Press, 1976.

Milgram, David L., and Rosenfeld, Azriel. "Object Detection in Infrared Images." In Leonard Bolc and Zenon Kulpa, eds. *Digital Image*

Processing Systems. Berlin: Springer-Verlag, 1981, pp. 228–353.

Miller, Stephan W. "A Compact Raster Format for Handling Spatial Data." *Technical Papers of the American Congress on Surveying and Mapping*, Fall 1980, paper no. CD–4–A.

Miranda, S. M. "Aspects of Data Security in General-purpose Data Base Management Systems." In *Proceedings of the 1980 Symposium on Security and Privacy, April 14–16, 1980, Oakland, California*. New York: Institute of Electrical and Electronic Engineers, 1980, pp. 46–58.

Mitchell, William B., and others. *GIRAS: A Geographic Information Retrieval and Analysis System for Handling Land Use and Land Cover Data*. Washington: U.S. Government Printing Office, U.S. Geological Survey Professional Paper no. 1059, 1977.

Moellering, Harold. "Designing Interactive Cartographic Systems Using the Concepts of Real and Virtual Maps." In *Proceedings of the International Symposium on Computer-Assisted Cartography, Auto-Carto VI, October 16–21, 1983*. Ottawa: Steering Committee of Auto-Carto Six, 1983, vol. II, pp. 53–64.

Moellering, Harold. "The Challenge of Developing a Set of National Digital Cartographic Data Standards for the United States." *Technical Papers of the American Congress on Surveying and Mapping*, 42nd Annual Meeting, 1982, pp. 201–12.

Moik, Johannes G. *Digital Processing of Remotely Sensed Images*. Washington: National Aeronautics and Space Administration, no. NASA SP–431, 1980.

Monmonier, Mark S. "Automated Techniques in Support of Planning for the National Atlas." *American Cartographer* 8 (1981): 161–68.

Monmonier, Mark S. "Cartography, Geographic Information, and Public Policy." *Journal of Geography in Higher Education* 6 (1982): 99–107.

Monmonier, Mark S. *Computer-assisted Cartography: Principles and Prospects*. Englewood Cliffs, N.J.: Prentice-Hall, 1982.

Monmonier, Mark S. "DIDS—A Defacto National Atlas." *Bulletin, Geography and Map Division, Special Libraries Association* no. 132 (June 1983): 2–7.

Monmonier, Mark S. "Map-Text Coordination in Geographic Writing." *Professional Geographer* 33 (1981): 406–12.

Monmonier, Mark S. "Private-sector Mapping of Pennsylvania: A Selective Cartographic History for 1870 to 1974." *Proceedings of the Pennsylvania Academy of Science* 55 (1981): 69–74.

Monmonier, Mark S. "Topographic Map Coverage of Pennsylvania: A

Study in Cartographic Evolution." *Proceedings of the Pennsylvania Academy of Science* 56 (1982): 61–66.

Montgomery, Bradley O. "The NAVSTAR Global Positioning System." *Professional Surveyor* 3, no. 5. (September/October 1983): 13–17.

Montouri, John S. "Image Scanner Technology." *Photogrammetric Engineering and Remote Sensing* 46 (1980): 49–61.

Mood, Fulmer. "The Rise of Official Statistical Cartography in Austria, Prussia, and the United States, 1855–1872." *Agricultural History* 20 (1946): 209–25.

Moore, A. D. "Electrostatics." *Scientific American* 226, no. 3 (March 1972): 46–58.

Moore, Lionel C. *Cartographic Scribing Materials, Instruments and Techniques.* Washington: American Congress on Surveying and Mapping, Technical Monograph no. CA–3, 1975.

Mourant, Ronald R., Lakshmanan, Raman, and Chantadisai, Roongrojn. "Visual Fatigue and Cathode-ray Tube Display Terminals." *Human Factors* 23 (1981): 529–40.

National Research Council, Panel on a Multipurpose Cadastre. *Procedures and Standards for a Multipurpose Cadastre.* Washington, D.C.: National Academy Press, 1982.

Neustadt, Richard M., Skall, Gregg P., and Hammer, Michael. "The Regulation of Electronic Publishing." *Federal Communications Law Journal* 33 (1981): 331–417.

Nicholson, N. L. "Review: *Atlas of Canada.*" *Cartographica* 18, no. 3 (Autumn 1981): 133–34.

"On-line with Sci-Tex." *Graphic Arts Buyer* 12, no. 5 (September–October 1981): 46–48.

Ormeling, F. J., Sr. "The Purpose and Use of National Atlases." In Barbara J. Gutsell, ed. *The Purpose and Use of National Atlases.* [Also *Cartographica,* monograph no. 23.] Toronto: University of Toronto Press, 1979, pp. 11–23.

Panel to Review the Report of the Federal Mapping Task Force on Mapping, Charting, Geodesy and Surveying, July 1973, Committee on Geodesy, Assembly of Mathematical and Physical Sciences, National Research Council. *Federal Surveying and Mapping: An Organizational Review.* Washington: National Academy Press, 1981.

Panshin, Daniel A. "What You Should Know about Loran-C Receivers." Marine Electronics Series, Oregon State University, Extension Marine Advisory Program, April 1979.

Paul, Charles K., and Mascarenhas, Adolfo C. "Remote Sensing in Development." *Science* 214 (1981): 139–45.

Pearson, Karen S. "The Nineteenth-Century Colour Revolution: Maps in Geographical Journals." *Imago Mundi* 32 (1980): 9–20.

Perry, Robert. "Personal Computers as the EE's Best Friend." *Electronics* 54, no. 8 (21 April 1981): 115–60.

Petersen, Timothy R. "SYMAP: A Study of the Recent History of Cartography and Institutional Communication in Computer-assisted Cartography." Master's thesis, Syracuse University, 1982.

Peterson, Ivars. "Computer Crime: Insecurity in Numbers." *Science News* 122 (1982): 12–14.

Peucker, Thomas K., and Chrisman, Nicholas. "Cartographic Data Structures." *American Cartographer* 2 (1975): 55–69.

Peuquet, Donna J. "An Examination of Techniques for Reformatting Digital Cartographic Data. Part 1: The Raster-to-Vector Process." *Cartographica* 18, no. 1 (Spring 1981): 34–48.

Peuquet, Donna J. "An Examination of Techniques for Reformatting Digital Cartographic Data. Part 2: The Vector-to-Raster Process." *Cartographica* 18, no. 3 (Autumn 1981): 21–33.

Phillips, J. W., Ransom, P. L., and Singleton, R. M. "On the Construction of Holograms and Halftone Pictures with an Ink Plotter." *Computer Graphics and Image Processing* 4 (1975): 200–208.

Place, John L. "The Land Use and Land Cover Map and Data Program of the U.S. Geological Survey: An Overview." *Remote Sensing of the Electromagnetic Spectrum* 4, no. 4 (October 1977): 1–9.

Pollack, Andrew. "Phone-Cable 'Hybrids' Seen." *New York Times* 5 November 1981, p. D2.

Poynter, Alan. "Computer-assisted Cartography—The Dilemma Facing British Producers of Derived Mapping in the 1980s." *Cartographic Journal* 20 (1983): 31–34.

Prado, Elias. "Voice Input for CAD/CAM." *Computer Graphics World* 6, no. 6 (June 1983): 111–13.

Quackenbush, Robert S., Jr., Lundahl, Arthur C., and Monsour, Edward. "Development of Photo Interpretation." In *Manual of Photographic Interpretation*. Washington: American Society of Photogrammetry, 1960, pp. 1–18.

Raisz, Erwin. "Charts of Historical Cartography." *Imago Mundi* 2 (1937): 9–16.

Raisz, Erwin. *Mapping the World*. London: Abelard-Schuman, 1956.

Rein, Barry D. "Protecting the Content of a ROM—The Current State of the Law." *Computer Graphics News* 1, no. 4 (January/February 1982): 8.

Rhind, D. W., Evans, I. S., and Visvalingam, M. "Making a National Atlas of Population by Computer." *Cartographic Journal* 17 (1980): 3–11.

Richardson, D. E. "The Cruise Missile: A Strategic Weapon for the 1980s." *Electronics and Power* 23 (1977): 896–901.

Richardson, D. E. "Spy Satellites: Somebody Could Be Watching You." *Electronics and Power* 24 (1978): 573–76.

Riche, Martha Farnsworth. "Choosing 1980 Census Data Products." *American Demographics* 3, no. 11 (December 1981): 12–16.

Ristow, Walter W. "Lithography and Maps, 1796–1850." In David Woodward, ed. *Five Centuries of Map Printing*. Chicago: University of Chicago Press, 1975, pp. 77–112.

Robinson, Arthur H. *Early Thematic Mapping in the History of Cartography*. Chicago: University of Chicago Press, 1982.

Robinson, Arthur H. *Elements of Cartography,* 1st ed. New York: John Wiley and Sons, 1953.

Robinson, Arthur H. *Elements of Cartography,* 2nd ed. New York: John Wiley and Sons, 1960.

Robinson, Arthur H. "Mapmaking and Map Printing: The Evolution of a Working Relationship." In David Woodward, ed. *Five Centuries of Map Printing*. Chicago: University of Chicago Press, 1975, pp. 1–23.

Robinson, Arthur H., and Petchenik, Barbara Bartz. *The Nature of Maps*. Chicago: University of Chicago Press, 1976.

Robinson, Arthur H., Sale, Randall, and Morrison, Joel. *Elements of Cartography,* 4th ed. New York: John Wiley and Sons, 1978.

Robinson, Arthur L. "Problems with Ultraminiaturized Transistors." *Science* 208 (1980): 1246–49.

Roese, John A. "Stereoscopic Computer Graphics for Simulation and Modeling." *Computer Graphics* 13, no. 2 (August 1979): 41–47.

Roper, James E. "VDTs and Employee Productivity." *Editor and Publisher* 116, no. 43 (22 October 1983): 34.

Rowe, G. H. "The Doppler Satellite Positioning Technique." *New Zealand Surveyor* 29 (1981): 608–24.

Sachs, Samuel G. "Map Scribing on Plastic Sheets Versus Ink Drafting." *Professional Geographer* 4, no. 5 (September 1952): 11–14.

"Safety Rules Recognize New Hazard in Screens." *New Scientist* 91 (1981): 393.

Salichtchev, K. A., ed. "National Atlases." *Cartographica* monograph no. 4 (1972).

Sandbank, C. P., ed. *Optical Fibre Communication Systems*. Chichester: John Wiley and Sons, 1980.

Sassone, Peter G., and Schaffer, William A. *Cost-Benefit Analysis: A Handbook*. New York: Academic Press, 1978.

Scherr, Sol. *Electronic Displays*. New York: John Wiley and Sons, 1979.

Schmidt, Milton O., and Rayner, William Horace. *Fundamentals of Surveying,* 2nd ed. New York: D. Van Nostrand Co., 1978.

Sena, Michael L. "Computer Mapping for Publication." *Computer Graphics World* 6, no. 7 (July 1983): 68–76.

Senefelder, Alois. *The Invention of Lithography.* New York: Fuchs and Lang Manufacturing Co., 1911.

Shotwell, Robyn. "Networking Is the Promise—and Problem—of Computers." *Publishers Weekly* 220, no. 17 (23 October 1981): 19–28.

Shryock, Henry S., Siegel, Jacob S., and Associates. *The Methods and Materials of Demography.* Washington: U.S. Government Printing Office, 1971.

Sijmons, Koert. "Cartographic Applications of Digital Techniques." *ITC Journal* no. 1982–2 (1982): 131–38.

Silverstone, Stuart. "Interactive Videodisc Technology Applications." *Computer Graphics World* 6, no. 12 (December 1983): 59–62.

Skelton, R. A. *Maps: A Historical Survey of Their Study and Collecting.* Chicago: University of Chicago Press, 1972.

Smirnoff, Michael V. *Measurements for Engineering and Other Surveys.* Englewood Cliffs, N.J.: Prentice-Hall, 1962.

Smith, Anthony. *The Politics of Information: Problems of Policy in Modern Media.* London: Macmillan, 1978.

Smith, David Eugene. *History of Mathematics, Volume II: Special Topics of Elementary Mathematics.* New York: Ginn and Co., 1925; and Dover Publications, 1958.

Smith, Doyle G. "Raster Data Development in the National Mapping Division, U.S. Geological Survey." *Technical Papers of the American Congress on Surveying and Mapping,* Fall Technical Meeting, 1981, pp. 284–88.

Smith, Ralph Lee. "Processed Words: Plugging In to the New Technology." *Nation* 233 (1981): 313–15.

Smith, Richard Daniel. "Maps, Their Deterioration and Preservation." *Special Libraries* 63 (1972): 59–68.

Sorrell, P. E. "Training, Education and Employment in the Field of Computer-Assisted Cartography." In David Rhind and Tim Adams, eds. *Computers in Cartography.* London: British Cartographic Society, Special Publication no. 2, 1982, pp. 85–89.

Southard, R. B. "The Development of U.S. National Mapping Policy." *American Cartographer* 10 (1983): 5–15.

Starr, Chauncey, and Rudman, Richard. "Parameters of Technological Growth." *Science* 182 (1973): 358–64.

Stefanovic, P. "Digital Screening Techniques." *ITC Journal* no. 1982–2 (1982): 139–44.

Steinberg, Gerald. "The Ultimate Battleground: Weapons in Space." *Technology Review* 84, no. 1 (October 1981): 57–63.

Stephens, Malcolm J. "The USGS 1:2,000,000–scale Digital Data Base." *Technical Papers of the American Congress on Surveying and Mapping*, 40th Annual Meeting, 1980, pp. 436–43.

Stock, Rodney. "Introduction to Digital Computer Graphics for Video." *SMPTE Journal* 90 (1981): 1184–89.

Stover, Hank. "Graphics System Displays True 3D Image." *Mini-Micro Systems* 14, no. 12 (December 1981): 121–23.

Swain, Philip H., Siegel, Howard Jay, and Smith, Bradley W. "Contextual Classification of Multispectral Remote Sensing Data Using a Multiprocessor System." *IEEE Transactions on Geoscience and Remote Sensing* GE–18 (1980): 197–203.

Swatik, Donald S. "Nonimpact Printing." In A. D. Moore, ed. *Electrostatics and Its Applications*. New York: John Wiley and Sons, 1973, pp. 307–35.

Taranik, James V., and Settle, Mark. "Space Shuttle: A New Era in Terrestrial Remote Sensing." *Science* 214 (1981): 619–26.

Taylor, D. R. F. "The Cartographic Potential of Telidon." *Cartographica* 19, nos. 3/4 (Autumn/Winter 1982): 18–30.

Teicholz, Eric. "Geographic Information Systems: the ODYSSEY Project." *Journal of the Surveying and Mapping Division, American Society of Civil Engineers* 106 (November 1980): 119–35.

"Teletext: TV Gets Married to the Printed Word." *Broadcasting* 97, no. 8 (20 August 1979): 30–36.

Thompson, Don W. *Men and Meridians, Volume I: Prior to 1867*. Ottawa: Information Canada, 1966.

Thompson, Morris M. *Maps for America*. Washington: U.S. Government Printing Office, 1979.

Thompson, Morris M., and Gruner, Heinz. "Foundations of Photogrammetry." In *Manual of Photogrammetry*, 4th ed. Falls Church, Va.: American Society of Photogrammetry, 1980, pp. 1–36.

Thrower, Norman J. W. "Cadastral Survey and County Atlases of the United States." *Cartographic Journal* 9 (1972): 43–51.

Thrower, Norman J. W. *Maps and Man*. Englewood Cliffs, N.J.: Prentice-Hall, 1972.

Thrower, Norman J. W., and Jensen, John R. "The Orthophoto and Orthophotomap: Characteristics, Development and Application." *American Cartographer* 3 (1976): 39–56.

Toffler, Alvin. *The Third Wave*. New York: William Morrow and Co., 1980.

Townshend, John R. G. "The Spatial Resolving Power of Earth Resources Satellites." *Progress in Physical Geography* 5 (1981): 32–55.

Townshend, John R. G., ed. *Terrain Analysis and Remote Sensing*. London: George Allen and Unwin, 1981.

Treftz, Walter H. "An Introduction to Inertial Positioning As Applied to Control and Land Surveying." *Surveying and Mapping* 41 (1981): 59–67.

Tsai, Robert C. "High Data Density 4–color LCD System." *Information Display* no. 5–81 (May 1981): 3–6.

Tsipis, Kosta. "Cruise Missiles." *Scientific American* 236, no. 2 (February 1977): 20–29.

Ulaby, Fawwaz T., Bradley, Gerald A., and Dobson, Myron C. "Microwave Backscatter Dependence on Surface Roughness, Soil Moisture, and Soil Texture: Part II—Vegetation-Covered Soils." *IEEE Transactions on Geoscience Electronics* GE–17 (1979): 33–40.

U.S. Bureau of Land Management. *Manual of Instructions for the Survey of the Public Lands of the United States*. Washington: U.S. Government Printing Office, 1947.

U.S. Bureau of the Census. *Census of Agriculture, 1969*, Vol. 5, Special Reports, Part 15, Graphic Summary. Washington: U.S. Government Printing Office, 1973.

U.S. Bureau of the Census. *Data Access Descriptions: Geography*, DAD No. 33, May 1979.

U.S. Geological Survey. *Landsat Data Users Handbook*. Reston, Va.: U.S. Geological Survey, 1979.

Uotila, Urho A. "Useful Statistics for Land Surveyors." *Surveying and Mapping* 33 (1973): 67–77.

Urick, Robert J. *Principles of Underwater Sound for Engineers*, 2nd ed. New York: McGraw-Hill, 1975.

Uzanne, Octive. "The End of Books." *Scribner's Magazine* 16, no. 2 (August 1894): 221–31. Reprinted in *Printing History* 1, no. 2 (1979): 23–32.

Vail, Hollis. "The Home Computer Terminal: Transforming the Household of Tomorrow." *Futurist* 14, no. 6 (December 1980): 52–58.

"VDTs and the Fetus—Cause for Concern?" *Science News* 120 (1981): 377.

Wagner, G. R. "Decision Support Systems: The Real Substance," *Interface* 11, no. 2 (April 1981): 77–86.

Waldrop, M. Mitchell. "Imaging the Earth (I): The Troubled First Decade of Landsat." *Science* 215 (1982): 1600–1603.

Waldrop, M. Mitchell, "What Price Privatizing Landsat?" *Science* 219 (1983): 752–54.

Walsh, John. "Japan-U.S. Competition: Semiconductors Are the Key." *Science* 215 (1982): 825–29.

Ware, Willis H. *Public Policy Aspects for an Information Age*. Santa Monica, Calif.: The Rand Corporation, Rand Papers Series no. 5784, 1977.

Wastesson, Olof. "Computer Cartography and Geographic Information Systems in Sweden." In Olof Wastesson, Bengt Rystedt, and D. R. F. Taylor, eds. *Computer Cartography in Sweden*. [Also *Cartographica,* monograph no. 20.] Toronto: University of Toronto Press, 1977, pp. 7–9.

Waters, Peter. "Archival Methods of Treatment for Library Documents." In John C. Williams, ed. *Preservation of Papers and Textiles of Historic and Artistic Value II*. Washington: American Chemical Society, 1981, pp. 13–23.

Webster, Martin. "Buying a Minicomputer." *Journal of Geography in Higher Education* 4 (1980): 42–50.

Wedding, Donald K., and Ernsthausen, Roger E. "Large-Area Flat Panel Displays." *Computer Graphics World* 6, no. 4 (April 1983): 68–70.

Weiss, Harold. "Computer Security: An Overview." *Datamation* 20, no. 1 (January 1974): 42–47.

White, Marvin S., Jr. "A Geometrical Model for Error Detection and Correction." In *Proceedings of the International Symposium on Computer-Assisted Cartography, Auto-Carto III, January 16–20, 1978.* Falls Church, Va.: American Congress on Surveying and Mapping, 1979, pp. 439–56.

White, Robert M. "Disk-Storage Technology." *Scientific American* 243, no. 2 (August 1980): 138–48.

Whitmore, George D. *Advanced Surveying and Mapping*. Scranton, Pa.: International Textbook Co., 1949.

Whitted, Turner. "Some Recent Advances in Computer Graphics." *Science* 215 (1982): 767–74.

Wickland, Lynn R. "Evolution of Paper and Plastics as Related to Mapping." *Professional Geographer* 4, no. 5 (September 1952): 15–18.

Wicklein, John. *Electronic Nightmare: The New Communications and Freedom*. New York: Viking Press, 1981.

Wilford, John Noble. *The Mapmakers*. New York: Alfred A. Knopf, 1981.

Wilson, William K., and others. "The Effect of Magnesium Bicarbonate Solutions on Various Papers." In John C. Williams, ed. *Preservation of Papers and Textiles of Historic and Artistic Value II*. Washington: American Chemical Society, 1981, pp. 87–107.

Winkler, Captain. "On Sea Charts Formerly Used in the Marshall Islands, with Notices of the Navigation of These Islanders in General." In *Annual Report of the Board of Regents of the Smithsonian Institution for the Year Ending June 30, 1899*. Washington: Government Printing Office, 1901, pp. 487–508.

Withington, P. Tucker. "The Trusted Function in Secure Decentralized Processing." In *Proceedings of the 1980 Symposium on Security and Privacy, April 14–16, 1980, Oakland, California*. New York: Institute of Electrical and Electronic Engineers, 1980, pp. 67–79.

Wolf, Paul R., and Johnson, Steven D. "Trilateration with Short Range EDM Equipment and Comparison with Triangulation." *Surveying and Mapping* 34 (1974): 337–46.

Woodward, David, ed. *Five Centuries of Map Printing*. Chicago: University of Chicago Press, 1975.

Woodward, David, ed. "Introduction" to *Cartography and Art*. Chicago: University of Chicago Press, forthcoming.

Woodward, David. *The All-American Map: Wax Engraving and Its Influence on Cartography*. Chicago: University of Chicago Press, 1977.

Woodward, David. "The Study of the History of Cartography: A Suggested Framework." *American Cartographer* 1 (1974): 101–15.

Wright, John. "Surveying and Cartography of Natural Resources: Framework Surveys." *Progress in Physical Geography* 4 (1980): 588–95.

Wu, Sherman S. C. "Geometric Corrections of Side-Looking Radar Images." *Technical Papers of the American Society of Photogrammetry,* 49th Annual Meeting, 1983, pp. 354–64.

Yoeli, Pinhas. "Digital Terrain Models and Their Cartographic and Cartometric Utilisation." *Cartographic Journal* 20 (1983): 17–22.

Zimmerman, Edward K. "The Evolution of the Domestic Information Display System: Toward a Government Public Information Network." *Review of Public Data Use* 8 (1980): 60–81.

Zuckerman, Ed. "How Would the U.S. Survive a Nuclear War?" *Esquire* 97, no. 3 (March 1982): 37–46.

Glossary

ABSOLUTE ORIENTATION.   The orienting to ground control of a stereo model.

ACCELEROMETER.   An instrument for measuring acceleration, the rate of change in velocity.

ACTINICALLY OPAQUE.   Not transmitting those wavelengths of light to which a specified photographic emulsion is sensitive.

ACTIVE SENSING SYSTEM.   A sensing system that also is the source of the electromagnetic radiation for which it measures reflectance.

AEROTRIANGULATION.   The use of aerial photographs for measuring the angles in a network of triangles so known horizontal and vertical positions can be used to extend a survey to additional nodes of the network.

ALGORITHM.   A series of well-defined steps for solving a problem or controlling a manufacturing process; when coded for use with a digital computer, the algorithm becomes a *program*.

ALIDADE.   A sighting instrument (with or without a telescope) used for plotting directions on a plane table.

ALPHANUMERIC.   A combination of letters (A–Z) and numbers (0–9), as well as some punctuation symbols.

ANAGLYPH.   A diagram with two superimposed images printed in complementary colors and positioned for viewing stereoscopically through a pair of lenses, each with one of the same complementary colors.

ANALOG MAP.   A map that represents a spatial distribution by means of continuously variable physical quantities, as in a typical inked or printed map.

ANALOG PHOTOGRAMMETRY.   The technology of making measurements from aerial photographs through optical-mechanical instruments.

ANALYTICAL PHOTOGRAMMETRY.   The technology of making measurements from aerial photographs through mathematical and statistical methods.

ANEROID BAROMETER. An instrument for measuring atmospheric pressure with a small box or tube the flexible top of which rises or falls as the pressure of the surrounding air falls or rises.

ARC. A portion of a geodetic *triangulation* network, between two high-order stations.

AREA CARTOGRAM. A map with a projection in which each areal unit is distorted so that its size is proportional to its value for a transforming distribution, for example, population, number of electoral votes, or number of acres of farmland; the area cartogram provides a convenient base map for a choropleth map in which the size and visual impact of each area symbol is proportional to its area's importance, as measured by the transforming variable.

BACULUM. A staff or rod used in measuring or leveling by Roman engineers.

BALLISTIC MISSILE. A self-propelled, self-guided missile that will fall in a ballistic trajectory toward its target.

BAND. A cluster of contiguous wavelengths, often used as a single communication channel.

BANDWIDTH. The range of frequencies that can be accommodated by an antenna, cable, or other telecommunications medium.

BASELINE. A precisely established survey line, such as that part of a horizontal control network upon which distance estimates are based.

BATCH PROCESSING. A mode of computing whereby programs are collected in a queue for processing individually, without intervention by the user.

BEARING. Direction of a line measured as an acute angle from either north or south along a meridian.

BISOCIATIVE THINKING. Mode of thought whereby association of two or more related facts or concepts leads to the generation of a new idea.

BIT. A binary digit, or integer, that may have one of but two values, either 0 or 1.

BLOCK (CENSUS). A small area used for collecting aggregated census counts and commonly bounded by two mutually perpendicular sets of parallel street segments.

BLOCK FACE. One of the street segments bounding a block, usually represented by a range of odd or even address numbers.

BLOCK PRINTING. Printing with crude letters or pictures carved on wooden blocks.

BUFFER MEMORY. An area for the temporary storage of data, and useful, for instance, for accumulating plot instructions from a fast computer for use with a comparatively slower plotter.

CARDINAL DIRECTIONS. The four principal directions: north, south, east, and west.

CARTOGRAM. A map with areas or distances distorted to promote communication of a concept.

CELESTIAL NAVIGATION. Navigation whereby heavenly bodies are used to determine geographical location.

CENSUS TRACT. A socioeconomically homogeneous portion of an urban area ideally consisting of a group of contiguous blocks and used for reporting aggregated counts.

CHAIN. The representation in a geographic information system of a continuous common boundary between adjoining areal units; the chain consists of a list of points describing the shape of the boundary, and begins and ends at nodes, at which it joins other chains.

CHANNEL. A path for transmitting data or other communications.

CHANNEL SWITCHING. A telecommunications method that defines a single path through a network for a communications session between two points, in contrast to *packet switching,* whereby separate packets might traverse different routes and possibly arrive out of sequence.

CHECK BIT. An extra binary digit used for checking the accuracy of data transfer and set to 0 or 1 so that the sum of all bits in a word should always be, for instance, an even number.

CHIP. A small piece of silicon wafer containing the circuitry for a central processor or a memory unit.

CHOROBATES. A Roman leveling instrument based upon a long pole with a groove for water along the top.

CHOROPLETH MAP. A map with areal units colored or shaded so that the brightness of each symbolized area represents its numerical value for the distribution mapped.

CHRONOMETER. A portable, rugged, highly accurate and precise timepiece.

COAXIAL CABLE. A cable with many wires twisted together about a common axis.

COHERENT OPTICS. The branch of optics concerned with light energy of essentially the same wavelength.

COLD TYPE. Type not involving hot metal and usually set photographically.

COLOR INFRARED FILM. Film sensitive to the green, red, and near-infrared portions of the spectrum, and used to produce color prints with green shown as blue, red shown as green, and infrared shown as red.

COLOR COMPOSITE PRINT.   A photographic print generated using multispectral imagery so that areas of dominant green reflectance are shown as blue, areas of dominant red reflectance are shown as green, and areas of dominant infrared reflectance are shown as red.

COM (COMPUTER-OUTPUT-MICROFILM) RECORDER.   A high-resolution display device for writing pictures or text on a roll of film about 35 mm or 70 mm wide for enlargement or viewing with a microfilm reader.

COMPARATOR.   In photogrammetry, an optical-mechanical instrument for the precise measurement of positions on aerial photographs.

COMPOSITE POSITIVE.   A positive film image containing all map elements to be printed with a specific ink.

COMPUTER SERVICE BUREAU.   A firm that provides one or a range of computing services, such as plotting or specialized data processing.

CONTACT EXPOSURE.   The photographic transfer of an image on transparent film in direct contact with a photosensitive emulsion.

CONTOUR INTERVAL.   The difference in elevation between adjacent contour lines.

CONTOUR LINE.   A line connecting points at a constant elevation on a surface.

CONTROL NETWORK.   An areally extensive network of points linked by measured distances or angles and for which elevations or horizontal positions are known, relative to a specific *datum* and stated levels of precision and accuracy.

CONTROL SURVEYING.   Branch of surveying concerned with the highly precise measurement of horizontal or vertical position relative to established reference systems.

COPYIST.   A person who reproduces text or pictures, including maps, by hand copying from the original or another copy.

CORE.   The main memory of a computer.

CROSS-STAFF.   An ancient and medieval instrument used in navigation for measuring the elevation angles of heavenly bodies or gunnery targets; now obsolete.

CRT DISPLAY.   A cathode-ray-tube display whereby a picture is written on the phosphorescent inner surface of a vacuum tube by a beam of electrons.

CRUISE MISSILE.   A pilotless, self-guided, self-propelled aircraft, launched from land, sea, or air and able to cruise at a low altitude to avoid detection by radar.

CURSOR. A movable screen character on a visual display unit, or a movable, hand-held cross-hair device used to indicate positions on a digitizer.

CYAN. The color that results from either adding blue light to green light or subtracting red light from white light.

DATA BASE MANAGEMENT SYSTEM (DBMS). A software system for managing a data base by providing for storage, retrieval, analysis, display, and security.

DATA STRUCTURE. The organization of data in computer memory, especially the linkages or pointers among related data objects.

DATUM. A standard reference system or level for the precise measurement of horizontal or vertical position.

DATUM PLANE. A plane for referencing elevation, or a coordinate system for referencing horizontal location.

DEAD RECKONING. A method of navigating by traveling an assumed distance and direction from the previous location.

DECISION SUPPORT SYSTEM (DSS). A computer data retrieval, analysis, and display system designed for easy use in aiding managers in making decisions.

DECLINATION OF THE SUN. The angular distance of the Sun above the equator of an imaginary sphere, the poles of which are the apparent centers of motion of the heavenly bodies, which turn about in the sky during the night.

DEFAULT VALUE. A value or option used by a program if no other value or option is specified by the user.

DEGREES OF FREEDOM. The number of independent quantities required to determine fully the values of all variables describing the state of a system.

DEVELOPABLE SURFACE. A flattenable surface onto which positions on the globe may be projected for representation on a map.

DIFFRACTION. The bending of light rays upon passage from one medium into another, for example, from air into water.

DIGITAL COMPUTER. A computer in which numbers and characters are stored and operated upon as discrete numbers, with a specific and finite number of digits or bits allocated to each.

DIGITAL ELEVATION MODEL (DEM). A terrain file of elevations at the intersections on a uniform grid.

DIGITAL MAP. Digital cartographic data that may be read by a computer for numerical analysis or visual display.

DIGITAL TABLET. A flat data-entry device used with a stylus for indicating and recording the plane coordinates of point locations.

DIGITAL TERRAIN MODEL (DTM). A digital representation of the land surface, such as a grid of elevations or a set of point-coordinate lists representing elevation contours.

DIGITIZER. A data-entry device used to measure and record the plane coordinates of points.

DIME FILE. A Dual-Independent-Map-Encoding file in which linear segments are represented by both the unique reference numbers of the bounding nodes and the unique area unit numbers of the co-bounding polygons.

DIMECO FILE. A DIME file for the United States by county unit.

DIOPTER. An optical instrument used in ancient Greece for leveling and measuring angles.

DISC STORAGE. The recording of data on the magnetic surface of a spinning disc for rapid storage and retrieval.

DISTRIBUTED DATA BASE. A data base with various parts stored on two or more computers linked through a telecommunications network.

DOPPLER SHIFT. The apparent alteration in frequency for a wave pattern when the source is moving relative to the observation point.

DOT-DISTRIBUTION MAP. A map in which each discrete dot represents a set number of the objects comprising the distribution portrayed on the map.

DOT-MATRIX PRINTER. A printer that displays pictures or graphic characters as a fine-grained matrix of open cells and dots that might vary in size and, possibly, color.

DRUM PLOTTER. A computer-controlled drawing device in which the pen may be raised or lowered as well as advanced back and forth in the Y direction along a track below which the paper is moved in the X direction by rotating the drum on which it rests.

EASTING. That coordinate used to measure distance east of the origin in a rectilinear system of plane coordinates.

EDGE AVERAGING. The averaging of reflectance values in remotely sensed data when the ground area represented by a pixel includes two or more different land covers.

EDGE ENHANCEMENT. A process used to sharpen the contrast between adjoining features in an image.

ELECTROMAGNETIC PULSE (EMP). The burst of electromagnetic radiation produced by a thermonuclear explosion above or in the upper part of the atmosphere.

ELECTROMAGNETIC RADIATION. Energy propagated through space or material as an advancing electric and magnetic field.

ELECTRONIC DISTANCE MEASUREMENT (EDM). The determination of distance with a device that measures the round-trip time between source and reflector of a pulsed electromagnetic signal.

ELECTRONIC MAIL. The forwarding of messages in a computer/tele-communications system.

ELECTROSTATIC PRINTER. A printer in which an image is formed on a sheet of paper by small charges that attract a dark powder, or toner.

ELEVATION GRID. A data set representing a land surface for which elevation has been sampled at the intersection points on a uniform grid.

ELLIPSOID. A surface formed by rotating an ellipse about one of its two axes.

EMULSION. The thin photosensitive coating on a sheet of photographic film or paper.

EPHEMERIS. A publication listing the calculated, expected positions of observable heavenly bodies on various days at specific times.

EQUATION OF TIME. A mathematical function of time of year that provides a value which may be added (or subtracted) to mean solar time to yield apparent solar time, which varies from mean time throughout the year because of the eccentricity of the Earth's orbit and the inclination of the Sun to the plane of that orbit.

EQUATOR. The great circle formed by the intersection of the Earth and a plane through the center of the Earth perpendicular to the axis of rotation.

ERGONOMICS. The psychological study of the relationships between the work environment and human productivity and satisfaction.

ERROR ELLIPSE. An elliptic symbol representing the estimated directional variation of locational error in a control network.

ETCH. A plate of metal or other material coated with a resistant layer that is scratched or removed photographically so that the image element can be emphasized by eroding the exposed area with an acid solution.

ETCH-AND-STRIP. A transparent film material coated with a photosensitive, strippable, opaque layer so that, for example, polygon boundaries can be exposed and weakened during development, thereby permitting the easy removal from the polygon's interior of the opaque coating.

FAR-INFRARED. The range of not-quite-visible electromagnetic radiation, with wavelengths slightly longer than visible red.

FIBER OPTICS. The transmission of coded signals as pulses of light through long, thin, flexible fibers of glass.

FIXED STATION. A node in a control survey network for which the exact location is known, for example, through celestial observation or a higher-order survey.

FLATBED PLOTTER. A flat display device on which an image is drawn by shifting a pen in the Y direction along a track mounted on a gantry that moves in the X direction.

FLIGHT HEIGHT. The distance between the ground and a camera mounted in an aircraft and used to photograph the landscape below.

FLOPPY DISC. An inexpensive, thin flexible sheet coated with magnetic material for the storage and retrieval of data on a disc drive similar in principle to those used for rapid "hard-disc" storage.

FOCAL LENGTH. The distance in a camera between the optical center of the lens and the intersection of the optic axis with the film plane.

GEODESY. The mathematical science concerned with the size and shape of the Earth and with precision surveys of the Earth's surface.

GEODIMETER. An electronic-optical instrument for measuring distance; see *electronic distance measurement*.

GEOGRAPHIC INFORMATION SYSTEM (GIS). A computer-based system for the storage and retrieval—and usually the analysis and display as well—of two- or three-dimensional geographic data.

GEOID. A generalized but nonetheless complex three-dimensional figure of the Earth intended to represent mean sea level extended through the continents as well as across the oceans.

GLITCH. A usually minor but erratic and worrisome mistake in a data file.

GNOMON. The inclined elevated pin that casts a shadow on the face of a sundial.

GNOMONIC PROJECTION. A projection covering less than a hemisphere and on which great circles are represented by straight lines.

GRAPHICACY. The communication skill based on graphic marks and symbols in the same sense that literacy is based upon written words, articulacy upon spoken words, and numeracy upon numbers.

GRAVER. A sharp, pointed or broad-tipped cutting tool used in engraving or scribing.

GRAY LEVEL. A numerical measure of grayness on a scale from, for example, 0 for white to 100 for solid black.

GREAT-CIRCLE ROUTE. The shortest-distance route over a spherical surface between two points.

GROMA. An ancient instrument for determining right angles and level lines with the aid of a plumb bob.

GROUND SWATH. The linear belt of land scanned by the sensing system of an orbiting Earth-observation satellite.

GROUND TRUTH. Ground data obtained to guide the computer-assisted interpretation of remotely sensed data.

GUNTER'S CHAIN. A 100-link chain 66 feet [20.1 m] long used by early land surveyors in the United States.

GYROSCOPE. A device that consists of a disc or wheel-like rotor spinning rapidly about its axis and that tends to resist changes in the direction of the axis in accord with Newton's Second Law of Motion.

HALFTONE. An image formed by a fine-grained uniform grid of dots of variable size, typically on or transferred from a photomechanical printing surface.

HARD COPY. A more or less permanent image, as on paper or film, for example.

HARDENING. The strengthening of a computer or telecommunications network by replacing or supplementing microcircuits susceptible to damage by an electromagnetic pulse.

HARDWARE. The computing machinery, in contrast to the data and programs.

HOLLERITH CARD. A punched-card, invented in 1888 by Herman Hollerith, on which characters and numbers are coded according to the vertical configurations of holes arranged in columns.

HOLOGRAM. An image recorded as an optical interference pattern that yields a true three-dimensional image when placed in a beam of coherent light.

HORIZONTAL ACCURACY. The reliability of positions in a horizontal plane, that is, the reliability of latitude/longitude or plane coordinates.

HORIZONTAL CONTROL. A referencing system concerned solely with horizontal position, not with elevation.

INERTIAL POSITIONING. The estimation of relative location by means of accelerometers able to record relative displacements in the vertical and two horizontal directions from a point of origin.

INFRARED. 1) Near-infrared, or reflected infrared: the portion of the electromagnetic spectrum with wavelengths slightly longer than those of the red portion of the visible band, 2) Far-infrared or thermal infrared: the portion of the electromagnetic spectrum associated with heat radiation.

INFRARED SENSING SYSTEM. A sensing system able to measure and record thermal infrared (or heat) radiation.

INK-JET PRINTER. A computer-controlled device that produces pictures or characters by carefully directing a stream of sprayed ink droplets.

INSTANTANEOUS FIELD OF VIEW (IFOV). The normally circular field of view sensed by an airborne or satellite scanner for a single pixel.

INTAGLIO ENGRAVING.   The cutting of linear image elements below the surface of a printing plate to form cavities or troughs that hold ink for transfer to the paper during pressing.

INTERFACE.   An electronic translator that permits the undistorted transmission of data between two pieces of computing, storage, data-entry, or display equipment.

INTERSECTION.   A triangulation method whereby the scaled location of a point can be reconstructed on a survey plot from the field measurements of the opposite side of the triangle and its two adjoining angles.

JOYSTICK.   A positioning device that can be shifted freely in two dimensions to move a cursor or symbol in any direction on the screen of a VDU.

KEY.   A digital code used to identify a data object, or record, for retrieval.

KEYPAD.   A small, hand-held keyboard, usually with no more than 16 keys.

LAND SURVEYING.   Measuring to determine the position of property boundaries, as described in a deed, or to develop a legal description of boundaries already marked on the landscape.

LANDSAT.   One of a series of Earth-observation satellites in a near-polar precessing orbit designed to cover almost the entire Earth, with repeated coverage every 18 days; Landsat-1 was launched by the National Aeronautics and Space Administration on July 23, 1972.

LASER.   A device generating a beam of electromagnetic energy of the same wavelength or near so, in or not far from the visible band.

LATITUDE.   Angular distance north or south of the equator.

LAYOVER.   On a radar image, the displacement of the top of a landscape feature relative to its base, for example, when a vertical cliff appears inclined less than 90 degrees.

LEAST-SQUARES ADJUSTMENT.   A trial-and-error procedure for fitting a curve or polygon to a set of points so as to minimize the sum of the squared deviations of the curve or polygon from the points.

LETTERPRESS.   A method of printing with the inked image elements raised above the background areas on the printing plate; also called *relief printing*.

LEVEL LINES.   The straight-line segments along which a leveling survey is conducted.

LEVELING.   The referencing of elevations to a vertical datum usually by measuring the vertical difference in elevation between successive points along a traverse.

LIBELLA.   A frame with a plumb line attached for determining a hori-

zontal line as in the plane perpendicular to the plumb line; used by the Romans for leveling.

LIGHT PEN.   A pointing device used with a CRT to detect the time of passage of the electron beam so that the time elapsed since the start of a scan can be used to estimate the location of a point on the screen.

LIGHT VALVE.   An electro-optical device whereby a voltage, current, magnetic or electric field, or electron beam can control the amount of light transmitted.

LIGHTHEAD.   A small projector carried in place of a pen by a flatbed plotter so that lines may be drawn by a moving beam of light and that characters or symbols may be exposed where needed on a sheet of photosensitive film.

LIMELIGHT.   A light source, once employed in theaters, whereby a bright light is emitted by a small block of limestone heated intensely by an oxyhydrogen torch.

LINE OF POSITION.   On a Loran chart, a hyperbolic line showing a series of possible locations relative to a pair of master/slave stations.

LINE PRINTER.   A computer-controlled device that prints text or symbols one line at a time.

LINEAR ARRAY DETECTOR.   An antenna or sensor whose elements are uniformly spaced along a straight line.

LITHOGRAPHY.   A printing process based upon a flat plate on which the image areas are made to accept ink and the non-image areas not to accept ink; also called *planar printing*.

LONGITUDE.   The angular distance east or west of a reference meridian, usually the Greenwich meridian.

LORAN.   A LOng RAnge Navigation system whereby position is fixed by measuring the difference in time of reception for pulsed signals transmitted from two or more fixed stations.

LORAN CHART.   A chart showing some locational frame-of-reference information as well as families of lines of position for pairs of Loran master/slave transmitters.

MAGNETIC NORTH.   The direction of the compass needle, which usually deviates from true north (along a meridian) and changes slowly with time.

MAINFRAME.   A large computer, as distinct from a minicomputer, microprocessor, and the like.

MAPSAT.   One of a series of proposed Earth-orbiting satellites with high-resolution sensors designed for the collection of detailed information about terrain and other Earth features.

MASK.   A partially opaque sheet of film to be placed above the unexposed emulsion in order to block light and thus permit the transfer only of selected parts of the other film image used in a contact exposure.

MEAN SEA LEVEL. The mean elevation of the surface of the sea, usually based upon hourly observations taken over a period of 19 years or longer.

MEAN TIME. Also called *mean solar time,* the time based on the rotation of the Earth relative to the mean sun, with each year consisting of 365¼ days of 86,400 seconds; apparent time differs from solar time because of the inclination of the Earth's axis to its path of revolution about the Sun.

MENU. A list of commands and options arranged on a grid so that the desired option can be selected by pointing to its cell with a light pen or stylus; also a list of numbered choices from which a selection is made by entering through a keyboard or keypad the corresponding number.

MERCATOR PROJECTION. A projection centered on the equator with evenly spaced meridians but increasingly widely spread parallels so that a straight line represents a line of constant direction, called a *loxodrome* or *rhumb line*.

MERIDIAN. One of an infinite number of great circle routes from the North Pole to the South Pole.

MICROPROCESSOR. A small computer, with a chip memory and central processor, and usually with interfaces to peripheral display, data-entry, or storage devices.

MICROSECOND. One-millionth of a second.

MICROWAVE. Electromagnetic radiation with a wavelength between approximately 0.1 and 100 cm, used for long-distance telecommunications and rapid cooking.

MICROWAVE SENSING SYSTEM. A passive microwave scanning system that forms an image by scanning its antenna beam.

MINICOMPUTER. A computer that is smaller than a mainframe computer but with more memory and built-in software than a microprocessor.

MULTISPECTRAL SCANNER (MSS) SYSTEM. A scanner that collects simultaneous imagery in each of several different spectral bands, such as the green, red, and reflected infrared bands.

MULTITEMPORAL IMAGERY. A data set covering a single set of pixels with remotely sensed imagery for different times of day, seasons, or years, organized to detect change or provide additional useful information for computer-assisted interpretation.

NADIR.   The point on the ground vertically below the observer or the camera; also called the *ground nadir.*

NADIR POINT.   The point on an aerial photograph pierced by a vertical line through the perspective center of the lens system.

NAUTICAL ALMANAC.   A book published annually listing astronomical data required for celestial navigation.

NEAR-INFRARED.   See *infrared.*

NEGATIVE SCRIBING.   Scratching or engraving lines on thin film with an opaque coating to produce a negative image that can be transferred photographically.

NOCTURNAL.   A sixteenth-century navigation instrument used for finding the time of night from angles between the Polar Star and other prominent heavenly bodies.

NORTH AMERICAN DATUM.   The horizontal-control datum of the United States, originally developed in 1927, and revised in the early 1980s.

NORTHING.   Distance north of the origin of a plane coordinate system.

OBLIQUE TRIANGLE.   A triangle without a right angle.

OFFSET LITHOGRAPHY.   Planar printing in which an image is transferred, or offset, from the inked press plate onto a rubber-coated blanket cylinder, from which it is offset, again, onto a sheet or web of paper.

OPEN-WINDOW NEGATIVE.   A film negative with open areas for which image elements might be transferred in a contact exposure, possibly with a tint screen added between negative and unexposed emulsion so that the open areas will be screened rather than solid.

OPERATING SYSTEM.   The software that controls the running of programs, the transfer of information between CPU and peripherals, and entry and display of data, so that the computer need receive but few specific instructions from its operator.

OPTICAL SCANNER.   A device for sensing and recording by grid cell the intensity of light reflected from a manuscript image or photograph mounted on a drum or flat bed.

ORTHOPHOTO.   An aerial photograph from which the effects of tilt and relief displacement have been removed to yield a planimetrically accurate map.

ORTHOPHOTOQUADRANGLE MAP.   A quadrangle-format map with a planimetrically accurate photographic image and some symbolized terrain and man-made features.

ORTHOPHOTOSCOPE.   A photomechanical instrument for producing orthophotographs.

PACKET.   A set of bits grouped for transmission as units in a data transmission network; may include address bits and error-check bits as well as message/data bits.

PACKET SWITCHING.   A data transmission system for the storing and rapid forwarding from one network node to another of addressed message/data packets; unlike with channel switching, separate packets in the same communication session might follow a variety of paths through the network and even arrive out of sequence.

PAGES.   A segment of a program or data file that may be transferred as a unit between main and external memory in a virtual memory computing system; the size of main memory thus appears far larger than its physical size.

PARALLAX DIFFERENCE.   On a pair of overlapping aerial photographs viewed in stereo, the difference in the distances separating comparable image points for terrain points at different elevations.

PARALLEL.   The circle, assuming a spherical Earth, of all points at the same angular distance north or south of the equator.

PARALLEL SAILING.   A method of navigation whereby a craft proceeds first to the latitude of its destination and then east or west to the appropriate longitude.

PARALLELEPIPED CLASSIFIER.   A classification method in remote sensing whereby a category is defined by a range of values for each of two or more spectral bands.

PARTICIPANT OBSERVATION.   A social science research method whereby the investigator establishes a close attachment to or membership in the group under observation.

PARTICLE-BEAM WEAPON.   A device for destroying enemy satellites or ground facilities and personnel with a concentrated, unidirectional flow of nuclear particles.

PASSIVE SENSING SYSTEM.   A sensing system that forms images by measuring and recording the reflectance of energy transmitted initially from some other source, such as the Sun.

PASSWORD.   A message used by a single user or group of users to gain access to a computing system or protected files.

PATTERN OSCILLATOR.   An electronic device for generating a periodically varying high-frequency electrical signal.

PEELCOAT.   Transparent film coated with a thin, strippable, opaque coating that can be peeled easily from the interior of a polygon after the boundary is cut mechanically, as with cut-and-strip peelcoat, or etched photographically, as with etch-and-strip peelcoat.

PEN PLOTTER.   A computer-controlled device for drawing lines on paper with a pen, on scribing film with a cutting tool, or on photosensitive film with a beam of light.

PERAMBULATOR. A wheelbarrow-like surveying instrument with a wheel and counter by which overland distance can be estimated from the number of revolutions of a wheel with a known circumference.

PERIPHERAL EQUIPMENT. Any device, for data entry, display, storage, or other functions, connected to a central processor.

PERSPECTIVE VIEW. The planar representation of a three-dimensional object with lines of projection converging to a central point, as the object might appear to the eye.

PHASE SHIFT. The difference in phase angle between the transmitted and reflected/received versions of the same periodic electronic signal.

PHOTO CELL. An electronic component that generates a small electrical current when exposed to light.

PHOTO-MECHANICAL. A system that involves both photographic and mechanical elements, as in etch-and-strip peelcoat, which must be exposed and developed as well as stripped by hand.

PHOTO-OFFSET LITHOGRAPHY. An offset lithographic printing process with press plates produced photographically from film negatives or camera-ready positive copy.

PHOTOGRAMMETRY. The art or science of obtaining measurements from photographs.

PHOTOGRAPHY. The art, science, or technology of obtaining or transferring images by means of light and light-sensitive materials.

PHOTOHEAD. See *lighthead*.

PICTURE DESCRIPTION INSTRUCTION (PDI). A specification of the position, orientation, size, color, and type of geometric figure or similar graphic element; an ordered sequence of appropriate picture description instructions can direct a microprocessor-controlled display device to reconstruct a coded, transmitted picture.

PILOT CHART. A chart containing meridians, parallels, coasts, navigation hazards, landmarks, and other information useful for marine navigation.

PIXEL. A picture element, usually a square or rectangular grid cell for which the land-cover attribute or band-specific reflectances are stored.

PLANAR. Flat, lying in a two-dimensional euclidean plane.

PLANE COORDINATES. Distances recorded relative to a pair of orthogonal measurement axes.

PLANE TABLE. A drawing board mounted on a tripod for surveying by constructing a network of triangles from angles drawn in a level, horizontal plane with a sighting device.

PLANIMETRIC. Measurements in a level, horizontal plane, as is usually represented by a flat large-scale map.

PLASMA PANEL. A display device with a gas in the narrow space between two transparent plates, each ruled with a set of closely spaced parallel conductors orthogonal to the other set to form a fine grid of cells. Any one of the cells can be illuminated by energizing the appropriate pair of nearly touching, intersecting wires to cause a gas discharge.

PLAT MAP. A scale diagram of property, subdivision, and political boundaries, as well as some physical features to provide a frame of reference; intended to record ownership of land, mineral claims, and other property rights.

PLATFORM. The aircraft, satellite, or other object upon which a remote sensor is mounted.

PLUMB BOB. A small weight at the end of a plumb line pulling the line in the direction of maximum gravitational attraction.

PLUMB LINE. A vertical line, in the direction of maximum gravitational attraction; not necessarily perpendicular to the geoid or coincident with a line from the point to the center of the Earth.

PORTOLANI (PORTOLAN CHARTS). A sailing chart showing lines of constant direction between ports or prominent landmarks.

PRIME MERIDIAN. A standard, widely accepted reference meridian from which longitude might be measured.

PRINCIPAL POINT. The perpendicular intersection of the optic axis of the camera with the focal, or film, plane.

PUSH-BROOM SCANNER. A satellite-mounted scanner for remote sensing, or a flatbed optical scanner capable of recording reflectance simultaneously for an entire scan line of pixels.

QUADRANT. One quarter of an area, typically divided by an intersecting meridian and parallel into northeast, southeast, southwest, and northwest quadrants.

QUANTITATIVE REVOLUTION IN GEOGRAPHY. The period during the late 1950s and 1960s when many academic geographers were captivated by the potential of statistics, probability theory, and quantitative modeling.

RADAR. A RAdio Detection And Ranging system, or technique for detecting and locating objects by measuring their reflection of radio-frequency radiation.

RADAR SENSING SYSTEM. A remote sensing system whereby terrain images are formed by measuring the extent to which the land surface scatters radio-frequency waves.

RADIAL DISPLACEMENT. The distortion on an aerial photograph of the relative planimetric positions of terrain features because of elevation differences and the central perspective of the photo; displacement is inward or outward from the center of the photograph.

RANDOM ACCESS. The trait of a data storage and retrieval device, such as a disc unit, whereby storage locations need not be accessed in a single sequence, as with magnetic tape.

RASTER. The row-by-row organization of the scan lines on the screen of a CRT display.

RASTER DATA. A gridded data structure, with an area divided into cells organized by row and usually by column as well.

REAL TIME. A mode of computing in which data are processed as they are received, as in industrial process-control computing and computer-assisted traffic regulation.

RECONNAISSANCE. A general exploration of the principal features of a locality or region.

RECTIFICATION. The reprojection of an aerial photograph to remove *tilt,* a deviation between the optic axis of the camera and a plumb line.

REFLECTANCE. The proportion of the radiant energy reflected by the surface upon which it is incident.

REFLECTED INFRARED. Electromagnetic radiation with wavelengths slightly longer than visible red.

REFRESH. Mode of operation whereby the electron beam of a CRT display rescans the screen 30 or more times per second to provide a changing image that does not flicker.

REGISTRATION. The horizontal alignment of colored symbols overprinted in different inks.

RELATIVE ORIENTATION. The alignment and adjustment of one of a pair of aerial photographs to the other photograph.

RELIEF DISPLACEMENT. See *radial displacement.*

REMOTE SENSING. The measurement of surface characteristics or imaging of the terrain from a distant aerial or satellite platform with an electronic or optical device for measuring or recording electromagnetic radiation.

REPEATER STATION. A device that receives and retransmits a signal, as in a microwave-relay transmission system or a slave station in a Loran system.

RESAMPLING. The estimation of reflectance values or attributes for a new pixel grid, with a resolution or orientation different from that of the original grid.

RESECTION. The graphical determination of position as the intersection of three or more lines of known direction from the point in question to known positions.

RESOLUTION.   The minimum distance between two distinct adjoining features, or, in remote sensing, the minimum size of a feature that can be detected by a sensing system.

RETRACEMENT SURVEY.   A survey undertaken to establish in the field the true line of an original boundary.

RETURN BEAM VIDICON (RBV) SYSTEM.   A sensing system in which an instantaneous image is formed in the sensor and then scanned by a television camera for transmission to a receiver station on Earth.

REVISION.   The act of changing map content to remove erroneous symbols and reflect change over time in the features portrayed.

RHUMB LINE.   A line of constant geographic direction—if the compass is corrected for magnetic declination, also a line of constant compass direction; also called a *loxodrome*.

RIGHT-READING.   An image not reversed—not a mirror or *wrong-reading* image.

ROM.   A Read Only Memory, usually a chip, from which stored programs or data may be read or executed, but on which other information may not be stored.

RUN-LENGTH ENCODING.   A compact data structure in which a scan line is represented only by the number of continuous runs of pixels with the same attribute and for each run, its length and the attribute.

SATELLITE POSITIONING.   A method for determining precise location based on the principle of the Doppler shift and several orbiting satellites, each of which broadcasts a radio signal that includes its current position.

SCAN LINE.   A row of a raster grid, to be printed on a single line of a raster display device or recorded in a single sweep of the mirror of a satellite-mounted multispectral scanner system, or a single addressing of the linear array detector of a pushbroom scanner.

SCAN PATH.   The path on the ground examined by one sweep of the scanner of an orbiting remote sensing satellite; the scan path is perpendicular to the trend of the ground swath and covers an area represented by one or several adjoining scan lines.

SCANNER.   A device that scans systematically an image on the ground and measures and records the radiation received; for example, an optical scanner or a multispectral scanning (MSS) system.

SCENE.   A set of remotely sensed data covering an approximately square area and formatted as a grid for convenient processing.

SCRIBECOAT.   A sheet of transparent film coated with an opaque surface that can be removed by a cutting tool to produce a negative image of cartographic lines and symbols.

SEA LEVEL DATUM OF 1929.   The "mean-sea-level" vertical reference surface established for the United States, to be replaced by a more accurate datum in the late 1980s.

SEASAT.   An Earth-observation satellite, equipped with a *SLAR* (side-looking airborne radar) sensing system, for collecting marine data.

SEMICONDUCTOR.   A crystal, or solid-state device, for controlling the flow of an electric current.

SENSOR.   An instrument that measures and records electromagnetic radiation.

SEPARATION.   A photographic image, positive or negative, containing all features of the same type or group of all symbol elements to be printed with the same colored ink on a polychrome map.

SERVOMECHANISM.   A feedback control system that compares the actual and intended positions or states of a mechanical device and determines the appropriate correction to decrease the disparity.

SLAR (SIDE-LOOKING AIRBORNE RADAR).   An active radar sensing system that scans outward, away from the flight path of an aircraft; SLAR systems can penetrate haze and operate at night.

SLICE LEVEL.   A threshold value in image processing used to assign all pixels with a higher (or lower) value to the same category.

SMART TERMINAL.   A microprocessor that can communicate with a larger, host computer or operate independently, in *stand-alone* mode.

SMOOTHING ENHANCEMENT.   The simplification of an image by eliminating small isolated features and making boundaries and linear features less jagged.

SOFTWARE.   Programs and large, documented data files.

SOLAR TIME.   Time based on the rotation of the Earth relative to the Sun, with the Sun reaching its highest position in the sky at noon.

SONAR.   A SOund Navigation And Ranging system whereby submerged objects and the sea bottom are detected by measuring the time elapsed between the generation of underwater sound waves and the reception of the reflected waves.

SPACE SHUTTLE.   A human-operated space vehicle launched by a rocket to orbit the Earth and then return to Earth and land undamaged on a runway like a conventional airplane.

SPARSE MATRIX.   A two-dimensional array with comparatively few non-zero, or non-background, elements.

SPECTRAL SENSITIVITY.   Sensitivity to radiant energy, as a function of wavelength.

SPHEROID.   A three-dimensional figure formed by rotating an ellipse about its shorter axis to provide a representation of the form of the Earth more accurate than a sphere.

SPIKE.   A sharp, erratic, and probably erroneous deviation from a line.

STADIA ROD.   A graduated rod held vertically so that distance can be estimated by reading the length of rod subtended between two small lines engraved in the telescope of a transit.

STAND ALONE.   The independent operation of a device, such as a computing system.

STANDARD ERROR.   A measure of the precision of an estimate, computed as the square root of the mean deviation of individual observations from their best estimate(s).

STAR CHART.   A map of the sky indicating, for a particular time of the year, the relative positions of various stars.

STATION.   A point whose position is to be or has been determined.

STEREOPLOTTER.   An instrument for viewing a stereoscopic model and plotting a planimetric map or measuring elevation differences.

STEREOSCOPE.   A binocular instrument for viewing a stereoscopic model.

STEREOSCOPIC IMAGE.   The cognitive appearance of a three-dimensional object when a scene is viewed stereoscopically.

STEREOSCOPIC MODEL.   A three-dimensional model of the terrain represented on two overlapping, properly oriented aerial photographs.

STEREOSAT.   An Earth-orbiting satellite designed to take stereo images of the terrain by sensing reflectance both before and after it passes over an area.

STICK CHARTS.   Crude navigation maps made of sticks, reeds, and shells, such as those used by Polynesian islanders.

STICK-UP.   Referring to labels or symbols that can be affixed to cartographic artwork with a self-adhesive or paste backing.

STRATEGIC PLANNING.   Generally long-range defensive, and sometimes offensive, measures against an enemy's preparations for making war.

STRIPPING FILM.   A photographic material with an emulsion atop a thin layer that, when exposed and developed, can be cut, lifted, and affixed to another sheet of film; stripping film is particularly useful for preparing stick-up type for maps.

SUBTENSE BAR.   A horizontal bar of fixed length that serves as the base of a triangle for which the acute apex angle is measured to estimate distance between base and apex.

SUN-SYNCHRONOUS.   The synchronization of a satellite's orbit with solar time so that reflectance is always sensed at the same time of day, thereby eliminating sun angle as a significant source of variation.

SYNCHRONOUS SATELLITE.   A satellite with an orbit adjusted to rotate about the Earth once every 24 hours above the same point on the equator; also called a *geo-synchronous* satellite.

TACTICAL PLANNING.   Generally short-range offensive or defensive measures, such as the positioning of forces and armaments to threaten or engage an enemy in battle.

TELETEXT.   A one-way television message service either broadcast or carried by cable.

TELLUROMETER.   An electronic instrument for measuring distance through the phase comparison of reflected radio waves.

TERMINAL.   A device, usually with a keyboard and either a printer or video display, for communication with a computer.

TERRAIN COMPARISON.   The comparison of a digital terrain model with elevation differences sensed by a radar altimeter in order to determine current position and guide a cruise missile.

THEMATIC MAP.   A map portraying information on a specific topic such as geology, agriculture, or demography, rather than general, frame-of-reference geographic distributions such as a combination of political boundaries, hydrography, and transportation routes.

THEMATIC MAPPER.   A remote sensor, designed for Landsat-4, that records reflectance in the thermal infrared band as well as in the visible and near-infrared bands.

THEME.   A distribution, such as population density or prime farmland, portrayed on a thematic map; sometimes called a "coverage."

THEODOLITE.   An instrument for the precise measurement of horizontal and vertical angles.

THERMAL BAND.   That part of the electromagnetic spectrum, with wavelengths between approximately 3 and 20 micrometers, that produces a sensation of heat; also called the thermal infrared, or *far-infrared,* band.

TILT.   The angular deviation between a vertical line and the optic axis of an aerial photograph that is not truly vertical.

TIME-SHARE.   Referring to a computer that accommodates in main memory the programs of two or more users, each of which receives in turn, for very short periods of time, the attention of the central processor.

TONER.   A fine, black or colored, resinous powder used to form graphic images by electrostatic printers.

TOPOGRAPHIC.   Land-surface features, including relief, surface water, drainage lines, and principal man-made features.

TRACKER BALL.   A pointing device with which a user can control the position of a cursor on the screen of a graphics system by drawing the palm of the hand across a freely rotating, mounted ball.

TRAINING DATA.   In remote sensing interpretation, sets of pixels representing known types of land cover and selected to guide the classification of pixels elsewhere in the data set.

TRANSIT.   A surveying instrument for measuring horizontal and vertical angles, and including a telescope, possibly with stadia hairs for measuring distance with a *stadia rod*.

TRANSPONDER.   A receiver-transmitter, possibly carried by a satellite, that, when interrogated by a signal with a prearranged password, transmits a coded acknowledgment as well as other measured or stored data.

TRAVERSE.   A series of lengths and directions, as to define a boundary or route.

TRAVERSE BOARD.   A board with holes and pegs, and with separate portions used to record the direction and distance sailed.

TRIANGULATION.   A method of determining location and extending a survey through the measurement of a base line and selected angles and the application of the principles of trigonometry to estimate the lengths of other distances between nodes in the network.

TRIG TRAVERSE.   A course for which the navigator estimates distance and direction using the principles of trigonometry.

TRILATERATION.   The method of estimating the positions of nodes in a network of triangles by measuring only distances, not angles.

TURNKEY SYSTEM.   A computer system with all necessary documentation and interfaces and a straightforward command language; in principle, a turnkey system can be bought "off the shelf," installed with minimal delay, and made operational merely by "turning the key."

TYPE POSITIVE.   In cartographic production, a transparent sheet of film with positive images of all type, and possibly some associated symbols, to be printed in the same color.

UNIVERSAL TRANSVERSE MERCATOR (UTM) GRID.   A widely used plane-coordinate system, based upon the transverse Mercator projection and employing 60 zones worldwide, each extending from 80 degrees south to 84 degrees north and covering 6 degrees of longitude.

VACUUM FRAME.   A frame with a glass plate on one side and a rubber blanket on the other and from which the air is evacuated so that photographic materials are held flat and in close contact to minimize distortion in a *contact exposure*.

VDU (VISUAL DISPLAY UNIT).   A terminal that includes a screen similar to that of a television set, but not necessarily employing a *cathode ray tube* (CRT).

VECTOR. A mathematical entity with both direction and magnitude, for example, the line segment defined by the coordinates of two points in a plane.

VECTOR DATA. Data represented by lists of point coordinates.

VERNIER. A supplementary scale that permits the estimation of measurements more precise than those provided on the principal scale of an instrument.

VERTICAL AERIAL PHOTOGRAPH. An aerial photograph taken with the optic axis of the camera perpendicular to the horizontal plane, or nearly so, in contrast to an oblique aerial photograph.

VERTICAL BLANKING INTERVAL. The "blank-line" signals at the end of each scan of a television screen, intended to separate successive frames.

VERTICAL CONTROL. An accurate, precise set of reference points for elevation.

VIDEODISC. A platter about the size of a long-playing phonograph record but intended for the storage of television signals, usually on a magnetic coating; an optical disc is an opaque disc with the signal coded as short transparent arcs varying in length and width.

VIDEOTEX. A one- or two-way television message service, usually carried by cable or telephone lines.

VIEWDATA. A two-way television message service, usually broadcast or carried by cable.

VOICE DECODER. A device that can be programmed to convert selected spoken commands into electronic signals to guide the operation of such systems as interactive digitizing and editing stations.

WAX ENGRAVING. A now-obsolete method in which a reproducible cartographic image is formed by engraving lines, area patterns, and point symbols in a layer of wax formed on a metal plate.

WOODCUT. A printing plate, or portion of a plate, on which an image is formed by either carving away the non-image areas or incising the image elements.

WORD. A group of bits in computer memory that can be addressed as a unit.

WRONG-READING. A reversed image, as seen in a mirror or formed in ink on the printing plate or blanket for direct transfer to the paper.

ZENITH DISTANCE. The angular distance on the celestial sphere between the point directly above the observer to the celestial body in question.

ZOOM. A command or option of a graphics system for enlarging part of the image on the screen.

Index

Accessibility to geographic information by the public, 136–37, 142, 183, 184–85, 187

Accuracy: improvements in, 6–9; celestial navigation, 28; Global Positioning System, 31, 34, 39, 202; Loran, 37; cruise missile, 39, 42; digital maps for missile guidance systems, 40–41; land surveying, 53, 54, 55; electronic distance measurement, 54, 55, 56, 58; geodetic surveying, 56, 60–62; least-squares trilateration, 58; ellipsoid to represent geoid, 61–62; photogrammetry, 71, 205; inertial positioning, 73; instantaneous field of view, 98; pixel position, 103; hardcopy display hardware, 164–65

Aerial photography: use in surveying, 63–71 *passim*; geometry of single photo, 63–68, 70; stereocompilation, 67–69, 71; interpretation for land cover mapping, 85–86, 89; history of, 86–89; usc in ccnsuscs, 114

Aerial platforms: balloons, 86–87; carrier pigeons, 87; kites, 87; rockets, 87; aircraft, 87; satellites, 89, 97, 209–10; sun-synchronous orbit, 97–98; synchronous polar orbit, 97; synchronous equatorial orbit, 97, 109; Space Shuttle, 97

Aeronautical charts, 43–44

Air Force (U.S.), 55, 88

Air photos: geometry of, 63–68; ground nadir, 63–64; principal point, 63–64; relief displacement, 64–66, 94; parallax, 66–68; orthophoto, 70; photointerpretation, 85–86

American Geographical Society, 114

American Society of Photogrammetry, 88

Analog map: contrasted with digital map, 3–4; in navigation, 14, 15–16

Animated maps, 163

Arabs: preservation of surveying technology during the Middle Ages, 48, 49

Area cartogram. *See* Map projections

Army Corps of Engineers (U.S.), 79

ARPANET, 175, 176

Austria: early development of statistical maps, 113

Babbage, Charles, 159

Babylonians: contributions to development of surveying, 47, 52; early use of a census, 111–12

Baculum. *See* Cross-staff

Base mapping, 59, 61, 63, 81–84, 86, 108, 111, 183

Bean, Russell, 69

Becker, Ted, 193

ben Gerson, Levi, 49

Bergstrand, Erik, 54

DESIGNED BY IRVING PERKINS ASSOCIATES
COMPOSED BY GRAPHIC COMPOSITION, INC., ATHENS GEORGIA
MANUFACTURED BY CUSHING MALLOY, INC.
ANN ARBOR, MICHIGAN
TEXT IS SET IN TIMES ROMAN, DISPLAY LINES IN GILL SANS

ⱳ

Library of Congress Cataloging in Publication Data
Monmonier, Mark S.
Technological transition in cartography.
Bibliography: pp. 229–248.
Includes index.
1. Cartography. I. Title.
GA105.3.M66   1985   526   84-40499
ISBN 0-299-10070-7